Technical Design Solutions for Theatre
The Technical Brief Collection
Volume 1

Technical Design Solutions for Theatre
The Technical Brief Collection
Volume 1

Bronislaw J. Sammler, Don Harvey

Focal Press
Taylor & Francis Group

NEW YORK AND LONDON

First published 2002

This edition published 2013
by Focal Press
70 Blanchard Road, Suite 402, Burlington, MA 01803

Simultaneously published in the UK
by Focal Press
2 Park Square, Milton Park, Abingdon, Oxon OX14 4RN

Focal Press is an imprint of the Taylor & Francis Group, an informa business

Library of Congress Cataloging-in-Publication Data

ISBN 13: 978-0-240-80490-3 (pbk)

British Library Cataloguing-in-Publication Data
A catalogue record for this book is available from the British Library.

Table of Contents

RIGGING HARDWARE

RIGGING TECHNIQUES

SCENERY HARDWARE

SCENERY MATERIALS

SCENERY MECHANICS

SCENERY TOOLS

Sound

You'd think that reprinting the original *Technical Brief Collection* would be simple: send the publisher the original discs and call it a day. But we didn't feel completely comfortable doing that. We had winced far too often at the annoying typos that had marred the first printing, and now was our opportunity to fix them. Besides, we reasoned, if this was to be reprinted as the first volume of a two-volume set, the contents ought to be completely reset in the interests of graphic and typographic integrity.

From one point of view, we were asking for trouble — and we got it. We revisited the whole thing page by page, fixing typos, replacing slashed fractions with stacked fractions, applying new style sheets, adding trademark symbols, and touching up the original drawings and putting each one in a captioned frame, always being careful not to force changes in the original pagination. We proofread our corrections not just once but three times, our first and second passes inevitably revealing yet another handful of errata. It actually sounds like *less* work than it was. This "simple chore" took immeasurably more attention, effort, time, and energy than either of us anticipated or could really afford. Yet, we learned so much along the way that it was worth it — not from the work but from the material we were working with.

First, many of the articles in this ten-year-old publication describe techniques that are still clever, workable, and immediately useful. These articles certainly deserve republication, and there is no need to reinvent these particular wheels, though (embarrassingly) we're just as likely as anyone else to waste effort by overlooking them. A number of other articles have become springboards for development, inspiring technician/authors to refine the original by employing different materials or to apply it to a surprisingly different technical problem with great success. Finally, of course, changes in economy and technology (particularly in electronics, sound, and computers) have made some of the original articles obsolete. Yet, while the specific subject matter may no longer be timely, the ideas behind them certainly endure, and thus even these articles "stand the test." As a case in point, Larry Schwartz's "Computer-Assisted Lighting Design" describes a mainframe computer program for managing and distributing light-plot information. PCs have long since replaced mainframes, and modern stage lighting instruments have capabilities that were unheard of in the mid-1980s. Yet, Mr. Schwartz's article ought to be preserved and recirculated, for it is based on an understanding of the relationship between lighting design and database development — a relationship fundamental to today's most useful and popular PC-based programs.

In sum, as we worked we realized that each of the original articles merits republication pretty much as it appeared in the first printing, and we realized that *Technical Brief* has achieved new value. As a thrice-yearly publication it fulfills its early purpose as a means for sharing immediately useful techniques in theatre production. As a loose-leaf collection of articles, it has become required reading for technical production classes in programs around the world. And as a two-volume publication, it is also one of the few enduring, living archives of how this profession has changed and grown in the last quarter-century of the twentieth century.

So, we close with thanks and a toast: here's to the authors who gave us all Volumes I and II . . . and here's to those who will give us all Volume III!

— Ben Sammler and Don Harvey
New Haven, Connecticut
January 2002

Foreword

Most of us in the technical side of theatre acquire our practical knowledge from the theatres we work in and the people we work with. It has been my privilege to have worked with some very talented colleagues who work miracles with materials and come up with extremely imaginative solutions to technical and craft problems. However, very few of the techniques, formulas, etc., are recorded for future reference. While this lack of record sometimes results in a new discovery, more often the attempt to reconstruct a previously achieved solution results in a great deal of lost time. Technical theatre workers usually have an interest in sharing their ideas and work, but the priority is to get on with the next solution — not to describe the problem just solved. Even when they do attempt a description, many technicians find it difficult to write, draw, or verbalize their ideas so that others can understand. As a result, written technical ideas often seem overly simplified or terribly complex.

The editors and contributors to *Technical Brief* have provided us with a long-needed collection of information on how specific technical problems have been approached and solved.

Technical Brief articles are ideal reading for my attention span. They are short (one or so pages), contain very little editorializing (just the facts), and often I find an idea that makes me smile and say, "That's a neat idea. I'll need to remember that."

To me, the most valuable use of *Technical Brief* is not to copy how someone solved a particular problem, but to see the approach, materials, and techniques that were used and develop from them an idea of how I might approach the problems I need to solve.

I wish to commend those who have contributed to *Technical Brief* and want to encourage all technicians and craftspersons to assist in the continuation of *Technical Brief* by contributing in the future.

Technical Brief is an excellent bit of work. Congratulations!

— Robert R. Scales, Ridgefield, Connecticut
January 1992

The Yale *Technical Brief* articles published over the last ten years chronicle two increasingly exciting and sophisticated technologies: the first is, explicitly, that of technology applied to stage production, and the second is, implicitly, that of technology applied to publishing. The rapid development of desktop publishing has made communication more immediate, accessible, and attractive; the message has evolved with the medium to serve an ever-increasing and technically aware audience of students and professionals in performing arts production.

These reports — now numbering over one hundred — have become valuable "how-to" references for concise descriptions of technical applications and techniques. *Technical Brief* has more than met the original publication goal of serving as a forum for the transfer of simple technical information so that the body of knowledge grows instead of being continually rediscovered in isolation.

Actually, the first objective was to provide short research and writing exercises for graduate students in a technical writing and research course at the Yale School of Drama before they embarked upon lengthy thesis projects. After several years of reading research papers that seemed to become longer and longer with each class, I attempted to reduce the wordcount (and my time spent in reading) by turning a few assignments into really concise technical writing: one-page reports.

Using the NASA Tech Brief as an example, I made the first assignment, which resulted in a consensus complaint from the class that "Theatre doesn't have many high-tech things to write about." To illustrate the point that a *Technical Brief* could be a low-tech discussion of a less than sophisticated material, I wrote "1101 — Corrugated (Kraft) Cardboard as a Scenic Material." (This somewhat tongue-in-cheek, but factual, example was not preceded by 1100 previous reports: rather, the number "1101" identified "paper-based construction materials" according to a classification scheme that, at the time, seemed appropriate to categorize future reports. However, intricate cataloging was too onerous in those pre-computer days; consequently, things were just numbered serially from the example: 1101, 1102, etc.)

Over the next few years, the results of these class assignments were shared among those in the program at Yale, with an occasional *Technical Brief* article surviving to become an example for succeeding classes. In 1979, the technical faculty decided that we could pull together enough reports to make an initial distribution to the outside world; we hoped to inspire other submissions and reports that would allow us to create a quick, relatively informal and inexpensive publication for the sharing of information among theatres, practitioners, and schools.

With the initial effort under way and the responses favorable, Ben Sammler took on the administration of the technical production program and the publication of *Technical Brief*, when after twenty years of enduring the wet and icy winters of New Haven, I moved to Austin in 1980. Since then, I have been more of a cheerleader than a contributor, as my activities have more and more focused on arts management, sailing, and Austin's developing music industry.

I eagerly await each set of *Technical Brief*, and am always delighted by the simplicity and ingenuity of techniques and applications presented. Having referred many students and technicians to the three-ring binder that holds my copies of these reports (some of which have lost both their original and reinforced holes). I applaud this publication of a bound volume as a way of renewing and keeping these valuable reports together while making them generally accessible to future technicians.

I am proud to have been a member of the group that got this started more than ten years ago. Congratulations and many thanks to Ben Sammler, Don Harvey, the students, and the contributors who have developed and carried this publication forward.

— John Robert Hood, Austin, Texas
January 1992

Acknowledgments

Technical Brief represents a contemporary history of theatre technology — a field in which developments are only rarely documented and even more rarely communicated. The ideas it contains are those of the field's practitioners. We congratulate those authors — those who were shanghaied into writing as well as those who volunteered submissions — on the publication of their individual efforts in a milestone collection. Now, for the first time perhaps, many of them will come to understand what distinguished company they keep as *Technical Brief* authors.

— Don Harvey

From its inception in John Hood's classroom through its development from the drawing board to the computer to its current success as a medium for a profession that has few other opportunities for communication, *Technical Brief* has been and continues to be an exciting and gratifying project. Our profound thanks to all who have shared in the process.

— Ben Sammler

Technical Design Solutions for Theatre
The Technical Brief Collection
Volume 1

TECHNICAL BRIEF

Lighting

Hanging cardboards are easy-to-read quick-reference guides for hanging, circuiting, and focusing a light plot. Union electricians have used cardboards in a variety of formats for many years; the form presented here was derived from these early cardboards by Gilbert V. Hemsley, Jr. and his associates.

After the light plot and hook-up are completed, the lighting designer or the master electrician prepares the hanging cardboards. One cardboard is drafted for each lighting position. Cardboards are usually drafted from the electrician's point of view, from upstage looking downstage. The example in Figure 1 and the discussion below are for a lighting pipe. The basic form can be adapted for booms, lighting bays, and other special positions.

On a piece of white posterboard, 4" wide and as long as necessary, lay out the lighting pipe in $\frac{1}{2}$" scale. Add the dimmer, lamp, color numbers, and focus notes next to each instrument. You can include the circuit numbers on the cardboards if they are known, or the electricians can fill them in during the hang. Indicate the location of the instruments on the pipe by drawing a scale parallel to the pipe, or by noting the distance of each instrument from the center line. Use colored felt tip markers for easy reading. Standard colors are black for the dimension and focus notes, red for the color numbers, and blue for the dimmer numbers. Instrument type and wattage can be indicated by symbols or notes. Include the name of the production, the lighting position, and some means of correctly orienting the cardboard. If the cardboard is too long to be handled easily, score and fold it on the centerline.

Put an equipment list indicating each instrument and its color on the back of the cardboard as shown in Figure 2. Attaching an instrument symbol key and any hardware and cable notes is also a good idea.

Drafting time can be saved by cutting apart a copy of the light plot by position, pasting the pieces on cardboard, and adding the necessary information.

In the shop, the cardboards can be used as references when preparing cable bundles and packing individual positions. During the hang, the electrician assigned to a lighting position uses the cardboard for that position when collecting the necessary equipment, and hanging and circuiting the instruments. The cardboard fits conveniently into a back pocket if the electrician needs free hands. If the electricians select the circuit numbers, they record them on the cardboards. As positions are hung and circuited, the master electrician collects the cardboards and records the circuit numbers. The cardboards then serve as handy references when troubleshooting and focusing.

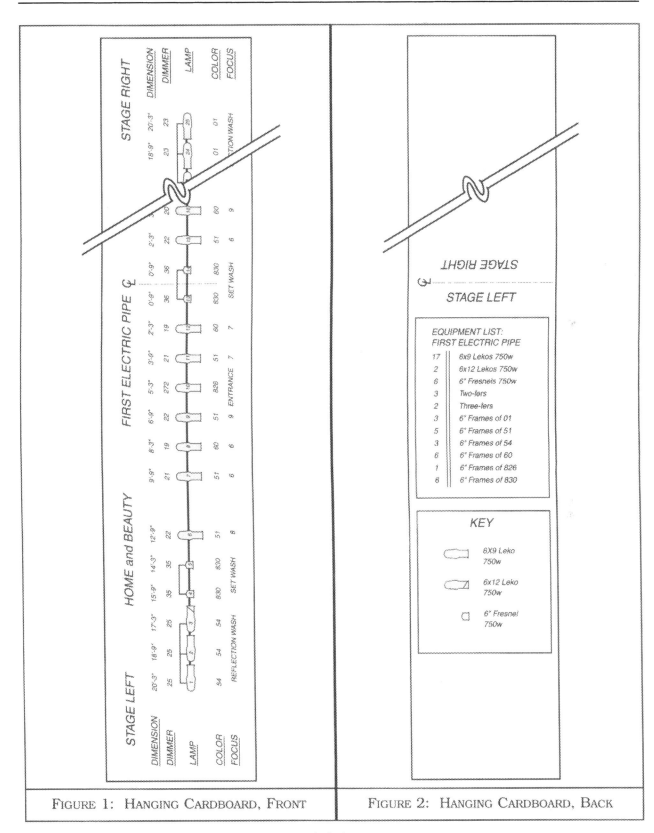

FIGURE 1: HANGING CARDBOARD, FRONT

FIGURE 2: HANGING CARDBOARD, BACK

A System to Facilitate Hanging Lights

Jerry Limoncelli

Rehanging an established plot can be facilitated by using the system described here. Suggested materials are index cards and sash cord. First cut the sash cord to the lengths of each electric as indicated by the light plot and mark the centerline of the plot on the cord. Next measure out from the centerline and affix an index card to the sash cord to represent each instrument. Write specific information about the related instrument — its type, wattage, color, and circuit number — on each card.

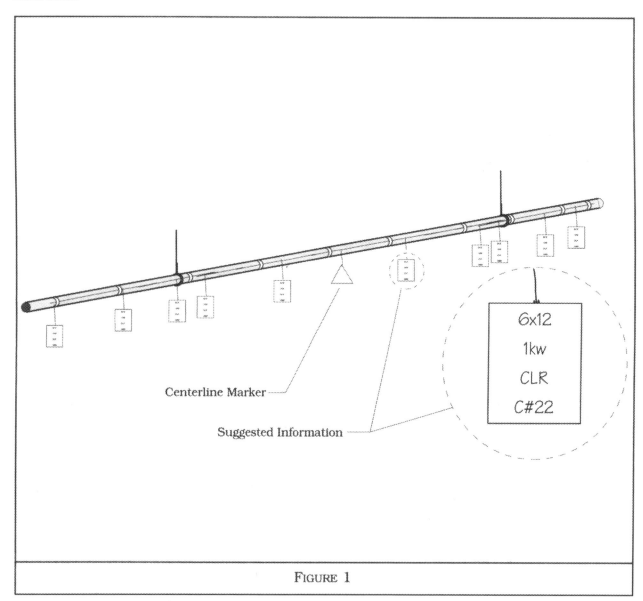

Centerline Marker

Suggested Information

6x12

1kw

CLR

C#22

FIGURE 1

At load-in, tape the prepared sash cords to the battens, working from the centerline toward each end. Figure 1 shows a batten labeled with this system. Even an inexperienced electrics crew can quickly and accurately hang lights by referring to the cards rather than the light plot.

Investing additional preparation time in the shop can save time during load-in.

❧❧❧

This article describes another way to facilitate the hanging of lights. The system is particularly useful during a quick turnaround such as a summer stock changeover. Materials needed include rolls of adding machine paper, black and red markers, a tape measure, and masking tape.

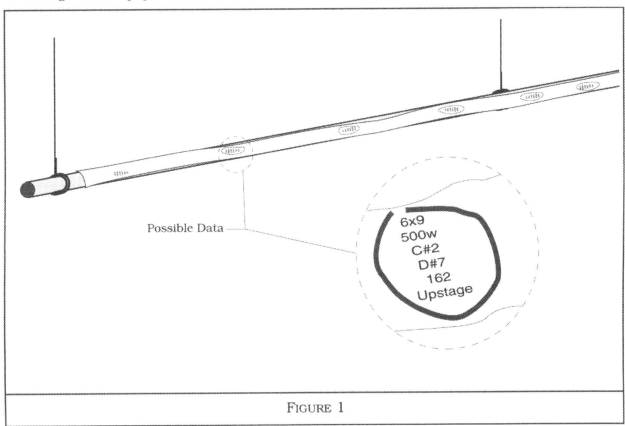

Possible Data

6x9
500w
C#2
D#7
162
Upstage

FIGURE 1

First, cut the adding machine paper to the lengths of the electrics. Second, find and mark the centerline. Third, measure out from the centerline in each direction and mark the hanging position of each instrument as indicated on the light plot. In a red-marker circle at each location, write the information illustrated in Figure 1: instrument type and size, lamp wattage, circuit number, dimmer number, color, and focus direction. To further facilitate the load-in, indicate the batten trim and weight on each strip, at the end closest to the rail.

At the beginning of load-in, after instruments have been struck from the battens from any previous production, tape the rolls of paper to each electric, aligning the centerline marks with the center of each electric. Starting at the center of each electric, you will be able to quickly and easily hang an entire show. After the lights have been hung and the cable has been run, the adding machine paper can be ripped down.

This system works particularly well on stripped battens.

❧❦❧❦

Computer-Assisted Lighting Design (C.A.L.D.) is a computer program developed at the Yale School of Drama by Joseph Barna. C.A.L.D. provides a means of storing lighting design data in one format and then retrieving it in the many formats used in stage lighting. Thus, it is especially useful in recording design changes. For example, making several changes — adding two 6x9 ellipsoidals to the system controlled by dimmer 3, reassigning the system controlled by dimmer 6 to dimmer 4, and adding two more systems of instruments to the design — normally means having to record those changes on several documents. By doing most of the paperwork, C.A.L.D. cuts the time involved by two-thirds. C.A.L.D. offers a variety of options too numerous to detail here, but this article describes the system's basic use and output.

INPUT

After drawing a light plot and making a list of instruments in any order, the designer enters the data into the program, working either by instrument or by position.

OUTPUT

Reports of the design's requirements are printed on a high-speed printer in several formats. Please note that the reports are not reproduced in scale here.

ELECTRICIAN'S REPORTS

Several reports give the electrician the data needed to hang the show. Each is described below.

> Hang: This report prints each position (*e.g.*, "Box Boom L") on a separate page, which can be used as an electrician's cardboard.

Instrument Number	Instrument Size	Type	Crt Watts	Number	Focus	Color	Notes	Done Date	Initials
1		6x22	1000		SL Cool Wash	60			
2		6x22	1000		SL Warm Wash	NC			
3		6x22	1000		SL Cool Wash	60			

> Equip: This report prints the equipment list.

Instrument Size	Description Type	Watts	Quantity
	— Par 64	1000	7
	— 6" Fresnel	500	11
	— 6x22	1000	6

Work: These reports are worksheets for keeping track of work to be done. Work may be dated and initialed as completed.

Position	Inst. No.	Instrument Description Size Type Watts	Work Notes	Done By / Date

Color: These are color-cut reports, which provide a total list of gels needed and a list of gel distribution by position.

Color	Frame Size or Instrument	Position	Done	Quantity Position	Total
04	7.5" x 7.5"	Beam		0	
		Box Boom Left		5	
		Box Boom Right		4	
		Corridor Boom		1	
		Electric 1		11	
		Electric 2		1	
		Toilet Boom 1		1	
					23

REHEARSAL REPORTS

These reports provide information for the lighting designer's use during rehearsals.

Schedule: This list of instruments is arranged by position.

Position	Inst. No.	Instrument Description Size Type Watts	Dimmer	Circuit	Focus	Color	Notes
Balc. Rail							
	1	8x11 Ellipsoidal 1000	Dim 1	CRT 246	SL	52	
	2	8x11 Ellipsoidal 1000	Dim 2	CRT 246	CL	52	
	3	8x11 Ellipsoidal 1000	Dim 3	CRT 250	SR	52	
Beam Lower							

Circuit: This list of instruments is arranged by circuit number.

Circuit	Dimmer	Position	Instrument Description			Focus	Color
			Size	Type	Watts		
10	72	Box Left 6	6x12	Ellipsoidal	750	Johnny on the Floor	NC
11	41	Box Left 7	6x16A	Ellipsoidal	1000	Table	L203
12	13	Box Left	6x9	Ellipsoidal	750	DSCL	NC
13	10	Box Left	6x9	Ellipsoidal	750	DSL	L203
15	8	Box Left	6x16	Ellipsoidal	750	DSR	L203
18	11	Box Left	6x16	Ellipsoidal	750	DSR	NC
21	30	Catwalk Left	6x16	Ellipsoidal	750	SR	L203
22	43	Catwalk Left	6x16	Ellipsoidal	750	SRC	L203
23	31	Catwalk Left	6x12	Ellipsoidal	750	SC	L203

Hookup: This list of instruments is arranged by dimmer number.

Dimmer	Dimmer Name	Position	Instrument Number	Instrument	Watts	Crt	Focus	Col
1		Box Boom Left	2	6x16 Ellipsoidal	750	18	DSRC	NC
2		Box Boom Left	7	6x9 Ellipsoidal	750	10	SL Rail	52
		Balcony Rail	1	8x11 Ellipsoidal	1000	241	SL	52
3		Balcony Rail	2	8x11 Ellipsoidal	1000	246	CL	52

Patch: This is the patch panel report, with the top-of-show patch described and room provided for the recording of re-plug information.

Dimmer	Top-of-Show Circuit	Replug After Cue
21	87	

In the future, C.A.L.D. will offer a graphics function that will print lighting symbols and light plots. The next step in the program's development will be to rewrite C.A.L.D. for use with personal computers, which could handle the data more efficiently. Presently, C.A.L.D. runs on an IBM 370 mainframe computer using the PL-1 language.

&a&a&a

Commercial-quality gobos can be made anywhere with some easily found materials and equipment. Disregarding the initial setup costs, making gobos is less expensive than buying them. The time it takes to make a batch depends on the complexity of the design, but three hours from tracing the design to fitting the finished product into holders is average.

MATERIALS AND EQUIPMENT

Muriatic Acid   Plastic or Bamboo Tongs
Krylon® Crystal Clear Spray   Tea Strainer or Window Screening
Carbon Paper   Rubber Gloves and Apron
Water   Splash-Proof Goggles
Acetone   Respirator with Organic Filter
Printers' Tin   Scrub Brush (acetone-resistant)
Photo-Developer Pans (at least 2)   Stylus or Scriber
X-acto® Knife   Plastic Jars with Screw-On Lids

Muriatic acid is available in gallons from swimming pool stores, and used sheets of printers' tin can be obtained free from most offset printers. Thus, the initial cost of setting up is reasonably low.

SAFETY NOTES

Acetone is highly flammable and its use creates the potential for explosion. Acetone is also a recognized health hazard. Always take appropriate precautions while using it.

1.   Work with plenty of ventilation and wear a respirator when using Krylon®, acid, or acetone.

2.   Protect the table top with plastic and absorbent paper.

3.   Keep plenty of water handy while cutting.

PREPARING THE METAL BLANKS

1.   Determine what size gobo will fit into your gobo holders.

2.   Cut the printers' tin to size with a paper cutter. These are your blanks.

3.   Spray all sides (including the edges) of the blanks with Krylon®. Apply at least three coats, allowing 15 minutes' drying time after the first coat and an hour after each subsequent coat. To save time later, lacquer a batch and keep them around.

4.   Position the blank in a gobo holder with the pattern to be traced on top of the blank and centered under the opening in the holder. The nature of the pattern has a direct bearing on the overall success of your efforts. When copying an already-cut gobo, the pattern will automatically be correct for etching. If the pattern is a designer's sketch or something else, however, make sure that the lines to be scribed on the blank will produce the desired effect when the acid cuts the scribed parts out. It is easy to make a mistake and have most or all of the gobo fall out because no metal is left to connect the center of the pattern to the edge of the gobo opening. If possible, make the pattern heavier in the middle to withstand the heat.

5.   Insert carbon paper between the pattern and the blank, and trace the pattern with a pencil.

6.   Remove the blank from the holder and scribe the outline of the design. The acid will cut through the metal at each scribed line. Don't scribe the holder opening!

CUTTING THE PATTERN

In this crucial step, the weaker the acid mixture, the slower the cut and the smoother the resulting edge will be. A good starting strength is one part acid to one part water. To strengthen the mixture in the pans after you start cutting, add a mixture of two parts acid to one part water. To weaken the pan mixture, add water. Watch the bubbles that form on the scribed lines of the blank. Bubbles should form slowly and not fizz to the surface of the acid bath. Fizzing will boil the lacquer away, ruining the gobo. Here are the steps:

1. Wearing goggles and a respirator, mix the 1:1 and 2:1 acid solutions in the plastic jars.

2. Set out as many pans as necessary to lay the blanks flat in a single layer. Stacking the blanks will hinder even cutting. Keep one pan unused for the acetone bath later.

3. Pour just enough 1:1 acid solution into the pans to cover a layer of blanks.

4. With the tongs, lay one blank in the solution scribed side up, and watch to see how fast the cutting happens. If bubbles form quickly, or if the solution boils, add water immediately! If nothing happens after a few minutes, add some of the 2:1 acid solution. Once the cutting speed is right, immerse all the blanks.

5. Stir the mixture gently and brush the bubbles away occasionally to keep the cutting even.

6. Remove the blanks before the lines are completely cut through so that the unwanted scraps come out of the bath along with the nearly cut design.

FINISHING

1. Drop the cut gobos into a bucket of clean water or rinse them under running water.

2. Wearing the rubber gloves, pour about $\frac{1}{2}$" of acetone into the unused pan.

3. Immerse the gobos, a few at a time, in the acetone and gently rub them with the brush. For delicate designs, use a cloth instead of a brush. The acetone neutralizes the acid; so immerse the gobos, don't just wipe them.

4. Repeat the acetone bath with clean acetone until no white lacquer residue remains on the gobo. Lacquer will burn in an instrument, leaving deposits on the lamp and lens, and creating fumes.

5. Trim any unwanted metal with the X-acto® knife.

CLEANUP

1. Screen any metal scraps out of the acid and dispose of the used chemicals properly.

2. Store unused acid solution in the plastic jars. Mark them clearly with the solution strength and the words "muriatic acid."

3. Wash the pans, tongs, brush, gloves, etc. with soap and water.

Old auto or truck inner tubes can easily be made into casings for a type of boom-base weight that offers all of the advantages of conventional sandbags while eliminating two disadvantages. Like conventional sandbags, inner-tube sandbags conform readily to the shape of jacks and tapered boom bases and give when kicked. Thus, both models are ideal for use in high-traffic areas.

Unlike conventional sandbags, however, inner-tube sandbags are rot-resistant and much less likely to rip open. Furthermore, they can be made much more quickly than canvas sandbags, and making them costs less than buying commercial canvas sandbags.

CONSTRUCTION

1. Cut a section of inner tube 3" longer than the desired finished length of the weight.

2. Cut four pieces of 1x2 long enough to span the width of the inner tube when it is folded back on itself.

3. Fold one end of the inner tube back on itself $1\frac{1}{2}$", sandwich the folded part between one pair of the 1x2s, and tack the completed sandwich together.

4. Carefully drill three or four $\frac{1}{4}$" holes (depending on the width of the inner tube) through the sandwich.

5. Securely bolt the sandwiched pieces together.

6. Perform steps three and four on the second end of the inner tube section.

7. Remove the tacks from the second end, fill the tube with sand, and bolt the second end shut.

CAUTIONS

Never use inner tube sandbags overhead or as counterweights in rigging systems.

❧❧❧❧

Easily Concealed Low-Profile Lighting Fixtures   *Jon Carlson*

The lighting designer's work is often hampered by the tightness of onstage lighting positions. Any set design — even a traditional box set with a ceiling — can be troublesome for the lighting designer when solid scenery completely masks critical onstage lighting positions.

Perhaps the best-known solution to this problem is to mount a three-inch Fresnel — an inky — behind a beam on the set or inside a piece of furniture. But lighting designers should be aware of three other solutions as well: the Lumiline® family of incandescent lamps, conventional showcase lamps, and compact fixtures like the Great American Stik-Up®. Any of these products can illuminate some dim corners and help balance your lighting. Moreover, they all run off standard 120V service and are completely dimmable. These features make them very useful in light boxes and other set pieces that require self-contained lighting.

LUMILINES® AND SHOWCASE LAMPS: SIMILARITIES

Lumilines® and showcase lamps have several features in common: both are powered by 120 volts AC, and are incandescent, dimmable, and tubular. Wired in parallel, several of either type of fixture can be placed end to end to achieve an even wash. In most cases, a mere two inches of solid scenery is necessary to mask them effectively. Thus, they can be installed behind headers and inside cornice molding, bookcases, and china cabinets.

The shape and size of Lumiline® and showcase lamps also make them ideal for use in very narrow light boxes. A light box fitted with Lumilines® and covered with a rear projection screen or white Plexiglas® can be framed out of 1x3 and work quite admirably as long as its front surface is covered with a diffusion medium that masks filaments and wiring.

Lumilines® and showcase lamps share one disadvantage: neither can be readily focused. Providing a reflective surface for the light to bounce off helps, but by no means does it solve the problem.

LUMILINES®

Designed for display cases, dressing room mirrors, and other places where the color of fluorescent light might not be appropriate, the Lumiline® lamp is a completely dimmable incandescent fixture with an unconventional disc base at either end. The base fits into a special socket called a Lumiline® cap. Normally, the caps are plugged into a special wall mount. For the theatre, the wall mounts represent added bulk and expense that can be avoided by soldering leads directly to the caps. Small wooden brackets cut from 1x3 will hold the lamps securely in position.

The Lumiline® is available in twelve-inch ($11\frac{3}{4}$") and eighteen-inch ($17\frac{3}{4}$") lengths. The twelve-inch unit comes in 40 watts; the eighteen-inch unit, in 30 or 60 watts. With all sizes there is a choice of three envelope finishes: clear, frosted, or white. The clear transmits the most lumens, and the white transmits the least. All Lumilines® have an overall diameter of one inch.

SHOWCASE LAMPS

As its name implies, the showcase lamp is used in museum and store showcases. Theatre technicians are also familiar with it because of its use on music stands for symphonies and orchestras. Unlike the Lumiline®, the showcase lamp has a medium screw base that makes wiring more straightforward. Showcase lamps are available in six-inch ($5\frac{5}{8}$") and twelve-inch ($11\frac{5}{8}$") lengths. The six-inch comes in 25, 40, and 60 watts, clear or frosted. The twelve inch comes in 40 and 75 watts, also clear or frosted.

STIK-UPS®

The Great American Market of Hollywood, CA, sells the Stik-Up®, a lightweight wire frame that supports a lamp housing and a reflector and mounts easily to scenery with gaffers' tape or wire. The manufacturer specifies the use of a Q/CL 100-watt quartz lamp.

Though the Stik-Up® has a reflector, its output is not particularly focusable. In addition, the Stik-Up® is somewhat expensive compared to the Lumiline® and showcase lamp. Nevertheless, it is very compact (3" x 3" x 4"), and Great American does sell special extension arms and clips for holding color media in place in front of the lamp.

AVAILABILITY

All of the low-profile fixtures discussed above are commercially available, and with a few phone calls can become part of any theatre's lighting inventory. If they are not stocked by your local lamp distributor, they can be obtained by special order. Discuss your needs with an electrical supply house and choose the fixture that suits your mounting needs.

Lumiline® and showcase lamps are manufactured by GE, GTE Sylvania, and North American Philips. The twelve-inch and eighteen-inch Lumilines® retail for $8.00 to $10.00 per lamp. The showcase lamps are less expensive than the Lumilines®. Lumiline® caps made by Leviton in Little Neck, NY, sell for about $.90 each. Call Leviton for the nearest distributor in your area. Stik-Ups® are available through theatrical supply houses and stage lighting companies that stock the Great American Market line of products. One Stik-Up® without accessories lists for $69.00; a set of three with all accessories will cost $289.00.

The problem presented was to produce a large number of fixtures for PAR-36-size ACL (aircraft landing) lamps. The fixtures would need to be focusable along both the horizontal and vertical axes. They should be durable, functional, inexpensive, easy to make, and should not look shop-built. They were to be mounted on an octagonal lighting truss made of welded square tube, and wired in series in groups of four to maintain the proper voltage.

Parts needed would include a barrel housing to hold the lamp, a yoke to provide for movement along the vertical axis, and a fastener to hold the unit to the truss and to provide for movement along the horizontal axis. It was immediately apparent that the material of choice for most of the basic components would be some sort of plastic. The right kind would be durable, easy to work with, and non-conductive.

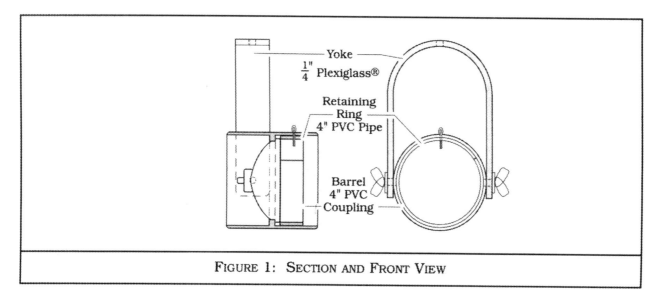

FIGURE 1: SECTION AND FRONT VIEW

For the barrel or housing of the lamp I tried various sizes and types of PVC pipe and fittings, happily discovering that a 100W, 28V #4591 ACL lamp nestled quite comfortably into a 4"-PVC coupling. As Figure 1 illustrates, a retaining ring of 4" PVC pipe was used to hold the lamp in place. For the yoke of the fixture, a $1\frac{1}{2}$" wide strip of $\frac{1}{4}$" Plexiglas® was heated and bent to the proper shape. Using $\frac{1}{4}$" round-head bolts and wing nuts for all fasteners allowed the units to be focused by hand — no wrenches required. In quantities of twenty or more, these fixtures should cost about $6.00 each and take 30 to 45 minutes to build.

BARREL CONSTRUCTION

1. Wrap a piece of heavy paper around the coupling and mark the placement of $\frac{17}{64}$" holes for the $\frac{1}{4}$" yoke bolts. Use the paper as a guide for marking the remainder of the couplings so that parts will be interchangeable.

2. Cut 1"-long sections of the PVC pipe. It is best to have a few spares and they will never be easier to make than right now. Turn the rings sideways and cut a slit through to the inside, just the size of the blade itself. Check for an easy fit into the coupling, and shave off some more if necessary. The lamps will burn out after about 28 hours and will need to be replaced. The rings should slide in easily but not fit too loosely.

3. To hold the rings in place, drill a $\frac{1}{8}$" hole through the top of the coupling and the ring, and insert a cotter pin. For uniformity, mark these holes using the paper guide used in Step 1.

YOKE CONSTRUCTION

1. Cut strips of $\frac{1}{4}$" Plexiglas® to the dimensions shown in Figure 2. I used white Plexiglas® to match the PVC, but color makes no difference if you will be painting them later. Be sure to sand all the corners and edges before you bend the yokes.

2. Using a jig to ensure exact duplication, drill $\frac{17}{64}$" holes as indicated.

3. Using two layers of $\frac{3}{4}$" plywood, make the jig illustrated in Figure 3. Both **A** and **B** will be used. Attach **B** to a scrap plywood base. It is best if the base has a very smooth surface. Trim an extra $\frac{1}{4}$" from the inside of **A** to allow space for the thickness of the Plexiglas®.

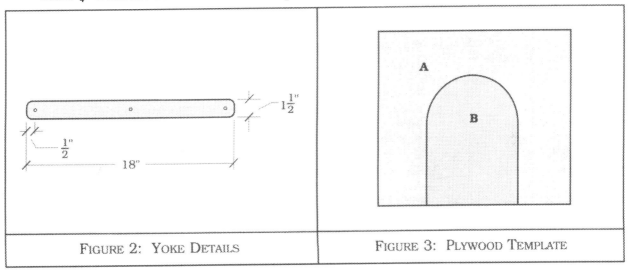

| FIGURE 2: YOKE DETAILS | FIGURE 3: PLYWOOD TEMPLATE |

4. Using a heat gun, heat the center 12" of the yoke until an easy plasticity is reached. It may be best to practice this on a piece of scrap if you've never done it before. Bend the yoke around your form, and then use **A** to hold it in place for a few seconds until it cools and becomes rigid.

ASSEMBLY

Assembly should be an easy and obvious matter by this point. Be sure to use wing nuts on the bolts in order to facilitate focusing. I used a $\frac{1}{4}$" hex head bolt to attach the fixture directly to the truss.

❧❧❧❧❧

Two Devices for Simplifying Cable Tracing

Steven A. Balk

Cable tracing can be a frustrating and time-consuming task. This article describes two inexpensive testing devices used by other industries that can make this job much easier and faster. The first device comprises two Progressive Electronics, Inc. (PEI) products: the Tracer®, model 77A; and the Inductive Amplifier®, model 200A. Together, these two units can be used to identify one particular pair of wires in a multi-conductor cable. The second testing device, the Tic Tracer®, manufactured by TIF Instruments, can be used to indicate the presence of voltage in on-line AC electrical equipment or cable.

IDENTIFYING CONDUCTOR PAIRS

Isolating one pair of conductors in a multi-conductor cable is difficult when color coding has been duplicated or ignored. Using a VOM to identify the pair can be very time-consuming because of the number of steps involved and the number of wire pair combinations you need to test for. A much easier and faster method makes use of the PEI products noted above.

The Tracer® generates an alternating signal that can be transmitted through a pair of wires. After connecting the leads of the Tracer® to one end of the wire pair to be traced, the technician moves the Inductive Amplifier® back and forth across all the conductor pairs at the opposite end of the bundle. During this process, the Tracer's® signal fades in and out, and the pair the technician wants to trace is the pair that provides the strongest signal.

Because the Inductive Amplifier® is capable of receiving the Tracer's® signal through the wire's insulation, it is unnecessary to strip conductors before tracing. Moreover, this pair of instruments can be used on either AC or DC circuits. Though the Tracer's® signal can be detected in live DC circuits, its signal resembles alternating current, and any AC conductors to be traced must first be taken off line. Nevertheless, using the Tracer® and the Inductive Amplifier® rather than a VOM, the technician can save a great deal of time since there is no need to short wires together, strip back leads, or test all the combinations of conductors one at a time.

LOCATING BREAKS IN LIVE AC CIRCUITS

It is not uncommon to string several shorter cables together in order to form one long AC circuit. If power does not reach the end of the cable, it is necessary to determine where the break is located. This, too, can be accomplished using a VOM, but, as with conductor-pair identification, it can be a very time-consuming process. Starting at the dimmer, the technician must check voltages at all connections. After finding a non-powered connection, the technician must discover whether the fault lies in the female connector, the male connector, or the cable itself.

TIF Instruments' Tic Tracer® offers a faster and easier method of locating such faults. The Tic Tracer® picks up the presence of the magnetic field created by an AC potential. The condition of a live AC circuit is indicated by the frequency of the audible tics the device generates. In the vicinity of a strong voltage the tics are close enough together to produce a steady tone; if there is a low voltage, the tics are farther apart.

Testing proceeds by powering the suspect line and testing each connection in turn, starting at the source. As long as there is a steady tone, the circuit is still sound. A steady tone on one side of a connection but not on the other indicates the presence of a fault.

FINAL NOTES

All three units described above are portable: the PEI units can be worn on a belt, and TIF's Tic Tracer® fits handily in a back pocket. In addition, their cost is extremely reasonable. Together the PEI devices cost approximately $60.00; TIF's Tic Tracer®, about $30.00.

Time is one of theatre technicians' most valuable assets. Whenever there is a complicated special effect with many control lines, or when large cable bundles are used to power remote lighting towers, there will most likely be wiring problems. By using either of the devices described here, these problems can be solved quickly and easily.

For more information, contact the manufacturers at the following addresses:

Progressive Electronics, Inc.
325 South El Dorado
Mesa, AZ 85202

TIF Instruments
9101 North W. 7th Ave.
Miami, FL 33150

TECHNICAL BRIEF

Lighting Effects

Simple Smoke

Michael D. Fain

Limited amounts of non-toxic smoke can be easily and inexpensively produced by heating small quantities of ammonium chloride (NH_4CL, sal ammoniac) in a heater cone or in an aluminum pan on a standard hot plate. Ammonium chloride can be found at most chemical supply stores for approximately $3.00 per pound. A heater cone is an inexpensive ($4.00) porcelain cone with a heating element wrapped around it. The cone holds a small quantity of powder (1 oz.) and has the advantage of heating up and cooling off more rapidly than a hot plate. A hot plate with a heating control allows more effective control over the duration of the smoke.

The white smoke produced rises slowly and can linger for 10 to 15 minutes in a space with little circulation. Another attractive feature of ammonium chloride smoke is that it doesn't leave a residue. No oil, wetness, or powder is left after it has dispersed. However, the airborne particles can be an irritant if the smoke is used in large quantities over extended periods of time. The smoke's chemical odor can be improved by adding a small quantity of cinnamon to the ammonium chloride.

It is necessary to test each application to determine the quantity of powder to use, the length of time it will take to produce smoke, the best device(s) to heat it in, and whether prevailing drafts will affect the direction of the smoke. A small quantity of powder heats and produces smoke faster than a large amount (4 to 6 minutes is average). After a portion of the powder begins to smoke, more can be added to sustain or increase the volume of smoke. For consistent results use fresh powder each time. When the heated powder turns dark brown it should be removed from the heat. More than one hot plate or cone can be used to increase the volume of smoke or to have it originate from more than one place. Many theatres have prevailing drafts on stage. Such currents should be noted during testing under full lighting and the sources of the smoke placed accordingly. If the smoke needs to disperse quickly, exhaust ventilation will be necessary. In many theatres smoke onstage will pour into the house, requiring exhaust ventilation in the house.

Two examples of the use of ammonium chloride smoke are as follows: first, a light smoke effect was needed to represent a small cooking fire. A hot plate was built into a fireproof box inside a fireplace. One ounce of ammonium chloride powder and cinnamon was poured into a pile in the center of the pan. Heated for 5 minutes, the powder smoked lightly for 8 to 10 minutes. The smoke rose slowly up and offstage, collected as a haze at about 12' after 6 or 7 minutes, and then dissipated within 3 or 4 minutes. The effect succeeded because it gave an indication of smoke without being obtrusive. Second, for a troll scene demanding a heavier smoke effect, one cup of powder was evenly spread to a depth of $\frac{1}{2}$" in a pan. After 5 minutes of heating, a light smoke developed for the transition and was quickly followed by billowing clouds for the trolls' entrance. The hot plate was placed stage left to take advantage of a left-to-right draft. The smoke was quickly drawn onto the stage where it remained for 10 to 15 minutes. The smoke revealed brilliant beams of light as the light illuminated the particles in the air.

SAFETY WARNING

Though non-toxic, the smoke generated by this method is widely recognized as a respiratory irritant. Each heater cone should be attended at all times by an operator with a fire extinguisher.

One common problem in using a fogger is noise. The machine's hissing sound is particularly undesirable in an intimate space, since it distracts the audience and also reminds them that technology is being used to create an effect.

Fortunately, the problem can be easily and quite inexpensively solved. C. Windsor Wheeler, Master Carpenter of the 1985 Champlain Shakespeare Festival in Burlington, VT, borrowed principles from the firearms industry to construct a silencer that solved the problem quite well.

Mr. Wheeler bored a series of uniformly spaced $\frac{3}{8}$" holes along the length of two pieces of schedule 40 PVC pipe, one measuring $\frac{3}{4}$" x $18\frac{1}{2}$", and the other, 3" x 18". See Figure 1 for hole placement.

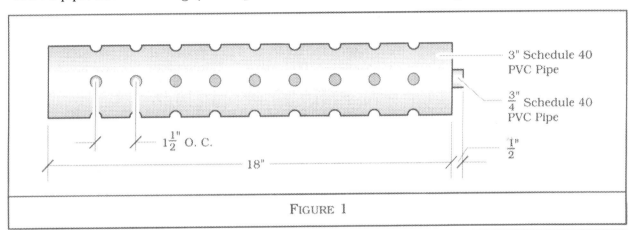

FIGURE 1

He then formed a cylinder of standard-density Styrofoam® to sleeve over the $\frac{3}{4}$" schedule 40 PVC pipe and inside the 3" schedule 40 PVC pipe. After coating the outside of the $\frac{3}{4}$" pipe and the foam cylinder with 3M's Fastbond 30®, he assembled the pipes and cylinder, offsetting the holes in the inner pipe by 45° as in Figure 2.

The $\frac{1}{2}$" long section of the $\frac{3}{4}$" pipe that protrudes out of the assembly slid easily but securely into the discharge opening of a Theatre Magic FR1 fog machine.

Not only does this device significantly reduce the noise produced by the machine, but, with the discharge end of the silencer fitted with a tee or other PVC connector, the machine is easily set up for satellite lines.

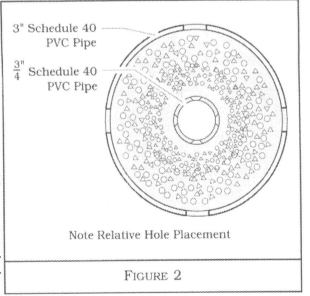

Note Relative Hole Placement

FIGURE 2

For a Yale Repertory Theatre production of *The Alchemist*, the set designer requested five practical chimneys that could smoke either independently or in conjunction with each other. The problem we faced was how to make one fogger serve as five. Master Electrician Donald W. Titus devised the fog manifold shown in Figure 1: a collection box fitted with five separately controlled PC fans that would deliver fog to desired locations through PVC piping. Turning one or more fans on pushed fog through the desired chimneys while pulling air backwards through any non-active chimneys.

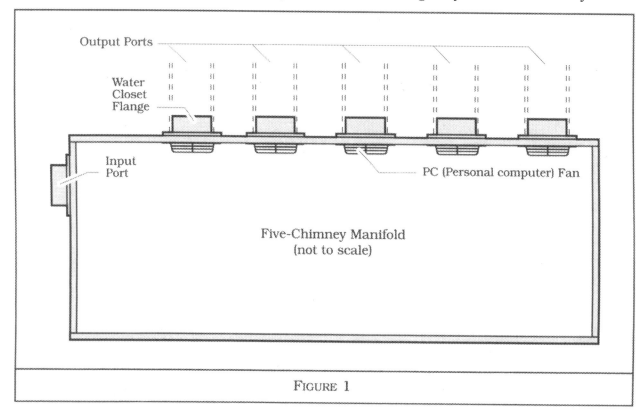

FIGURE 1

COMPONENTS

$\frac{1}{2}$" plywood Manifold Box with 3"-diameter input and output ports and a removable panel
PC Fans (available from Radio Shack)
Water Closet (toilet) Flanges with 3" mouths
3" schedule 40 PVC Pipe, Elbows, and Couplings
Dryer Hose and Duct Tape
Silicone Sealant

CONSTRUCTION

We built our box 18" high, 2' deep, and 3' long. The input port was cut into one end at a height that would align with a fog machine's nozzle when both fogger and box stood on the same surface. We cut five output ports into the top, enabling us to run our PVC pipe straight up, with the box taking their weight. A removable side panel allowed us to get to the fans inside the box as necessary. Assembly instructions follow.

1. Bolt the fans in place on the inside of the box, lining them up with the output holes, and putting the nuts inside the box.

2. Attach the water closet flanges to the outside of the box, again lining them up with the ports. If possible, use one set of bolts to join both fan and flange to the box. Otherwise, countersink the heads of the fans' bolts or use silicone sealant between flange and box. Whatever the approach, make the resulting seal airtight.

3. Wire the fans and run the wires out through a hole in the top or side. Caulk the hole.

4. Place the box in the most convenient position backstage and attach the PVC pipe. There is no need to cement the joints if the elbows and couplings fit snugly, particularly if the joints are duct taped. This practice also leaves the PVC pipe and connectors reusable.

5. Run the fogger into the box. Test each run for leaks and for fog-transit time by turning the fans on as they will be used in the show.

❧❧❧❧

As well as serving as technical support for the Fine Arts departments, the University Theatre of the University of Calgary Campus is a rental facility for all types of performances. One thing that these groups often want is fog.

These days, chemical fog machines are pretty reliable, and, with water-based juice, they're not even very messy. So fog is easy to do, right? Even dry-ice fog is no problem. In our theatre, however, these shows are often performed on a bare stage, and we had no method of getting the fog on stage other than setting up the outlets of the fog machines just inside the wings and letting it waft onstage. That would probably work except for the fact that in our theatre the prevailing winds blow the wrong direction, and the final effect was that I would either be waist deep in dry ice fog or suffocating in chemical fog backstage. Only 60% of the fog ever made it onstage, and it seemed like a waste, as well as being uncomfortable, for the little effect that it was having onstage.

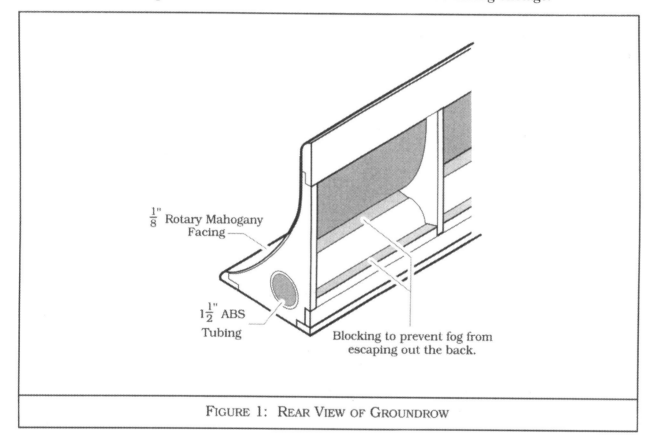

$\frac{1}{8}$" Rotary Mahogany
Facing

$1\frac{1}{2}$" ABS
Tubing

Blocking to prevent fog from
escaping out the back.

FIGURE 1: REAR VIEW OF GROUNDROW

When we decided that it was time to build a new groundrow for the stage, it occurred to me that here was a chance to try an experiment. Our groundrow hides the gap between the cyc and the stage floor, and gives us a place to hide strip lights for lighting the cyc. There would also be room inside for a bit of plumbing. The new groundrow (see Figures 1 and 2) is built in two 18' sections, replacing the old one, which was in five sections. Running through the length of each section is a piece of $1\frac{1}{2}$" schedule 40 ABS plastic pipe. In the center 10' of each section, slots cut in the pipe feed fog into a channel that leads to an opening in the groundrow's lower front edge. When the two sections of the groundrow are placed end to end, the pipe ends butt together. This allows fog to be fed in from either or both ends at once.

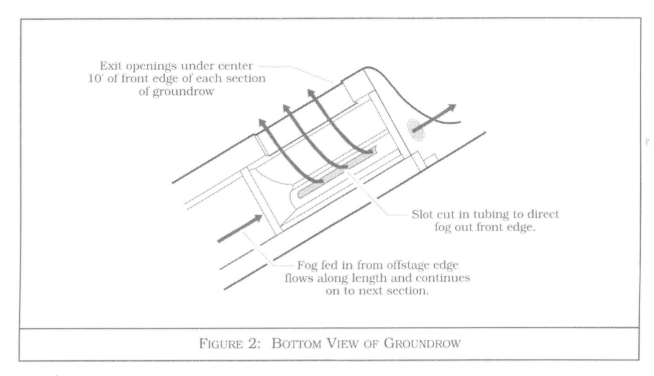

Exit openings under center
10' of front edge of each section
of groundrow

Slot cut in tubing to direct
fog out front edge.

Fog fed in from offstage edge
flows along length and continues
on to next section.

FIGURE 2: BOTTOM VIEW OF GROUNDROW

The $1\frac{1}{2}$" tubing with standard fittings found in any plumbing supply store fits perfectly onto the outlet nozzle of the Rosco fog machines that we use. A simple adapter from $1\frac{1}{2}$" to 4" allows easy changeover to the 4" flexible dryer hose we use on the dry ice machines. This rather inexpensive experiment (adding only the cost of the ABS pipe and fittings to the price of a piece of equipment we were building anyway) permits us to deliver the fog to center stage with a minimum of fuss.

NOTE

This article appeared first in the September 1989 *Alberta Section USITT Newsletter* and was submitted by the author for publication here.

❧❧❧❧❧

An effect of an exploding ship was desired for a production of *Peer Gynt*. We investigated the idea of using the Pyropak® airburst placed on a stand to create the effect. The Pyropak® airburst charge is a small quantity (1 teaspoon or less) of airburst flash powder contained in a 3" x 3" square of flash paper tied into a small bundle with an electric match inserted into the center. See Figure 1. The airburst charge leads are attached to the airburst transformer terminals, and the transformer is connected to the Pyropak® multi-channel controller.

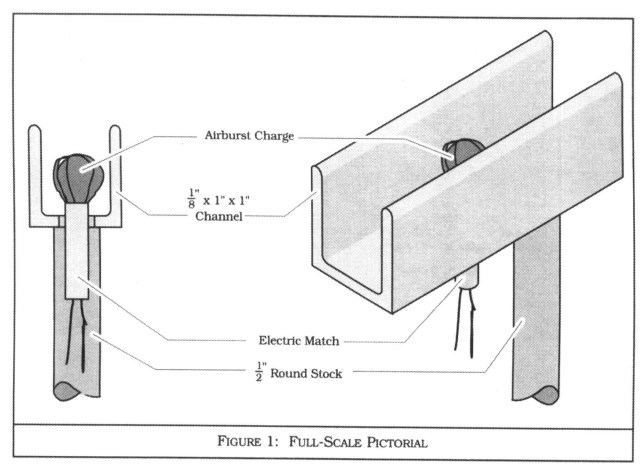

Airburst Charge

$\frac{1}{8}$" x 1" x 1" Channel

Electric Match

$\frac{1}{2}$" Round Stock

FIGURE 1: FULL-SCALE PICTORIAL

The ship, located 3' upstage of a white scrim and 4' downstage of a light blue cyc, was constructed of black tie line and flame-proofed muslin. The pyro stand was placed 1' upstage of the ship. Pyropak's® manufacturer, LunaTech, Inc., recommends keeping the stand's distance from flammable materials equal to the diameter of the burst, but our explosion's proximity to the ship and to drops made that impossible. No flammable materials were located on either side or above the area of the explosion, however, so we decided to change the shape of the explosion to a more nearly two-dimensional fan shape. See Figure 2.

We constructed a stand to direct the explosion as well as position it at the proper height. A $2\frac{1}{2}$" length of 1"-deep steel channel was welded to a 5' length of $\frac{1}{2}$" round steel rod that was threaded into a small boom base. A $\frac{1}{4}$" hole drilled near the center of the channel allowed the leads of an electric match (ignitor) to drop through and connect to the transformer terminal.

Through testing we determined that one cap (approximately $\frac{1}{2}$" teaspoon) of Pyropak® Regular Airburst Flash Powder® created a fan-shaped burst approximately 10' high and 20' wide.

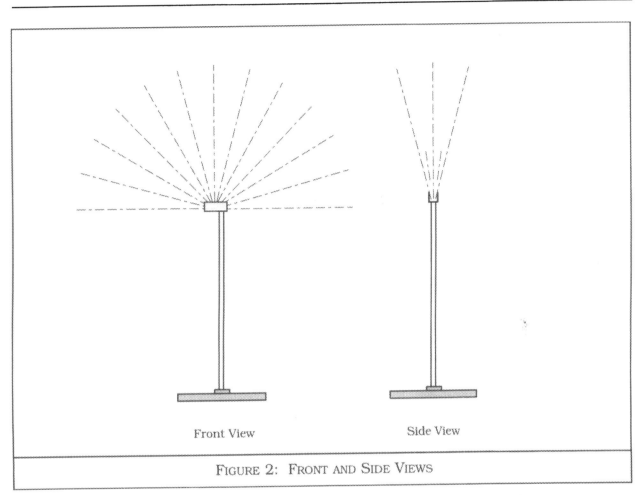

Front View   Side View

FIGURE 2: FRONT AND SIDE VIEWS

For more specific information about the Pyropak® airburst, contact the manufacturer:

LunaTech, Inc.
P.O. Box 2495
Huntsville, AL 35804
(205) 533-1487

❧❧❧

A Photographic Aid in the Preparation of
Scenic Projections

Jon Farley

When projected images must closely align with stage scenery, it becomes vital to know what the stage looks like from the projector's point of view. Careful preparation of artwork must compensate for the focal length of the lens, length of throw, and angle of the projector axis relative to the projection surface. In simple cases, these factors can be dealt with by drafting a distorted grid and scaling the artwork on it. In most cases, however, the preparation of such a grid is extremely difficult because of the complex geometry involved.

A simple technique useful in correcting artwork for projection geometry is the photographing of the stage with the scenic projector that will eventually be used to project the image. By doing this it is possible to create a print containing an image of the stage that is inherently corrected for projection angle and lens characteristics and exactly the same size as the final slide. This image can be traced directly in the preparation of hand-painted slides or enlarged to fit any scale in which the designer wishes to work. If a pre-distorted grid is desired, it can be easily produced by working from known vertical and horizontal lines in the image. Regardless of how the print is used, remember that the image is reversed and inverted so suitable corrections in artwork must be made.

The technique is simple, inexpensive, and can be performed by anyone with minimal darkroom experience. Slides that are correct the first time should be easy to produce. The following describes the process.

MATERIALS

Black and white photographic paper. Any moderately high contrast paper will do, but resin-coated paper is best. Kodabromide F®, Grade 3 would be a good choice.

Chemicals
> Kodak Dektol® Developer
> Kodak Indicator® Stop Bath
> Kodak Fixer

Three Processing Trays
Photographic Safe Light
Paper Cutter
Two Glass Slides or 35mm Glass Slide Mount depending on the projector
Focus Slide
Portable Floodlight (500W to 1000W quartz is best.)
$\frac{3}{4}$" masking tape

PROCEDURE

1. Line the set with masking tape. Outline important features with masking tape so they will be easily identifiable in the print. If the set is all white, a dark-colored tape should be used.

2. Install the projector. Place the focus slide in the projector and turn on the lamp. Aim the projector so its field covers the desired area and the image is in focus. With certain large lens projectors, it may be necessary to add a diaphragm to the front of the projector to effectively stop down the lens and bring the image to acceptable clarity. A suitable one can be made by cutting a 1" to 3" hole in a piece of matte board and placing it in the color frame of the lens.

3. Lock the projector down. From this time until the closing of the show, the projector should not be moved. If it is, accurate slide alignment will be very difficult to achieve.

4. Turn the projector off, allow it to cool, and remove the focus slide.

5. Set up the chemicals. The prints will be processed immediately after exposure so the trays should be convenient to the projector.

6. Darken the theatre. The theatre now becomes a photographic darkroom. All light except a safe light near the projector must be extinguished. Light leakage through doors must be stopped and exit signs temporarily turned off.

7. Prepare the photographic paper. Once the theatre is dark, open the package of paper. Cut a piece to exactly the size of the finished slide and return the remaining paper to its package. If desired, paper can be cut in advance and transported to the theatre in a light-proof container. For large-format projectors, sandwich the cut piece between two glass slides; for 35mm, place it in the glass slide mount. Place the mounted paper in the gate of the projector with the emulsion side facing the stage.

8. Expose the paper. Since most projectors do not have shutters, exposure is controlled by the amount of time the set is lit. This can best be accomplished by "painting" the set with a hand-held floodlight for a measured amount of time. Start with an exposure of 15 seconds and adjust as necessary. It is important that all areas of interest receive even doses of light and that the flood is never shown directly at the projector. A human figure included in the scene will give a sense of scale to the image.

9. Develop the paper. Remove the paper from the projector and process like any photographic print.

10. Restore light to the theatre and check the print. A few tries will be needed to get a satisfactory image. Adjust exposure time and concentration of light as needed to improve the results.

11. Document the print. Be sure to record exposure time, position of projector, and date on the back of the print. This will facilitate its later use.

ADDITIONAL COMMENTS

If desired, Kodak Ektamatic Stabilization Process Print Paper® can be used in this application. It has the advantage of a faster and simpler chemistry, which expedites the exposure process and eliminates some of the mess of the three open chemical baths. Its disadvantages are the expense of a processing machine (about $300), and the relatively short life of the resulting prints (about 6 to 8 months).

Do not attempt to use an auto-focusing projector. The auto-focus mechanism shines a separate light into the projector gate, which will fog photographic paper and thus ruin the print.

SOURCE

R. Duncan MacKenzie.

¿▲¿▲¿▲

A Fiber-Optic Star Drop

Billy Woods

The most commonly used technique to give the illusion of stars is low-wattage lamps, laboriously wired into a fairly bulky sky drop. This article describes a much lighter, more quickly built, and more durable sky drop, which uses fiber-optic filaments in place of both the lamps and the wires that connect them.

While high-grade fiber optics were developed for the communications field, Mitsubishi Rayon Company has developed a grade of less expensive fiber optics especially suited to theatre. Marketed under the trade name of Eska® in diameters of 0.1 to 3.0 millimeters, this durable plastic filament can be purchased in spools of up to 2.4 kilometers in length.

BUILDING A FIBER-OPTIC STAR DROP

In the earliest stage, building a fiber-optic star drop is no harder than building a low-wattage lamp star drop. The lighting designer maps out the location of the stars on a black drop with bits of masking tape. At the same time, the master electrician plots the best offstage location for the two or more 250W Fresnels or lekos that will be the sources of the starlight. The best locations are those that are well masked, are out of stage traffic patterns, and will result in the use of the shortest fiber-optic filaments. Once both the stars and the positions of the light sources have been located, assembly proceeds as follows:

1. Modify a 4x4 duplex box as follows. Spot-weld the open face of the box to a piece of sheet metal that will slide into the gel-frame holder of the light source. Screw romex connectors into the knockouts on the back of the box.

2. Poke individual strands of fiber-optic filament through the drop from the back, letting them extend about $\frac{1}{4}$" beyond the front and hot glue each filament to the drop at its point of penetration.

3. Group the filaments into "trunk lines" and run them along the back of the drop toward one of the starlight sources.

4. Cut the filaments to a generously long rough length.

5. Lay the trunk lines flat against the back of the drop and gaffers' tape them in place.

6. Where the trunk lines extend beyond the edges of the drop, heat-shrink or cable-wrap them into bundles that run to the light sources.

7. Trim the bundles to their final length and fit the ends with SO connectors to prevent crushing. Insert the ends into the romex connectors and tighten the connectors.

EMBELLISHMENTS

In the night sky stars vary in size, color, and intensity, all of which can be approximated by making changes at the starlight source instruments and/or at the star points. Varying the wattage of the lamps in the source instrument or putting a gel in the frame can add color variety. Variety can also be achieved by treating individual filament tips (the stars). An untreated tip has a 60° light output angle. Touching a tip with a soldering iron causes it to mushroom, increasing the output angle and making the star appear dimmer. Sanding a tip with very fine dry-wet sandpaper makes the star seem to sparkle when viewed from the house. Dipping tips in transparent lamp dye adds color dimension. A little creative experimentation may yield surprising results.

FINAL NOTES

The materials cost of a fiber-optics star drop is greater than that of a low-wattage-lamp star drop. A 2700-meter spool of 0.75mm Eska® costs approximately $900.00 wholesale. On the other hand, the savings in labor far outweighs the materials cost. Building a 400-star fiber-optic star drop took about 40 hours; building a low-wattage-lamp star drop of comparable size takes about 100 hours. Fiber-optic star drops also weigh less than low-wattage-lamp star drops, are extremely durable, and can be taken apart for reuse. They also take up fewer dimmers, allowing designers more control elsewhere, and simplify troubleshooting since there are only a few real light sources and no soldered connections.

SOURCE

Donald W. Titus.

A Television Lighting Effect

Michael Van Dyke

Many modern scripts call for a television onstage. Directors and lighting designers want to see the effect of the TV screen illumination on the actors and the scenery. The easiest way to achieve this effect is to plug in a working TV with the volume turned down. There are several disadvantages to a practical television, however. The intensity of the light coming from the screen cannot be adjusted beyond the limit of the set's brightness control. The effect cannot be faded in and out of the scene with the stage lights. And, unless a videotape is used, the effect is inconsistent since few stations broadcast the same episodes of the same shows night after night.

The effect of the light from a TV screen is an apparently random variation of intensity that changes smoothly through camera panning, fades, and the like, as well as abruptly through camera cuts.

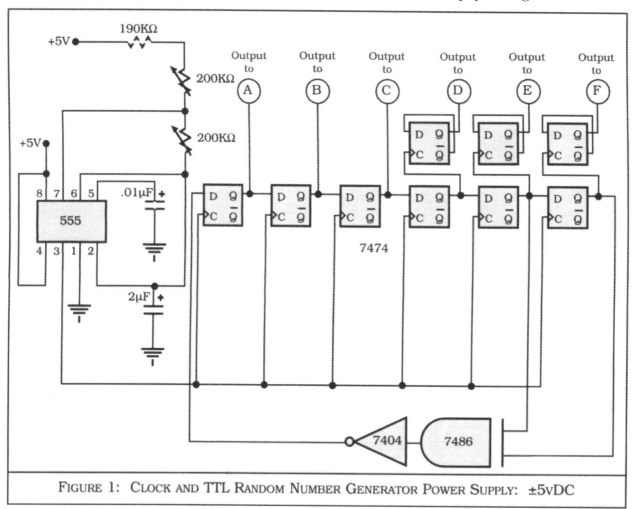

FIGURE 1: CLOCK AND TTL RANDOM NUMBER GENERATOR POWER SUPPLY: ±5vDC

To achieve this effect and solve the problems of using a working TV set, I devised an electronic circuit to vary randomly the controller output to an SCR dimmer to control a stage lighting instrument placed in a gutted TV set. See Figures 1 and 2.

The circuit is based on a pseudo-random number generator triggered by an electronic pulse maker (the "clock"). I chose a six-digit generator that gives a repeating sequence length of 63 distinct states, hence the term "pseudo"-random. The values of the first three digits are sent to an operational amplifier to be translated into voltage values that change immediately. The second three dig-

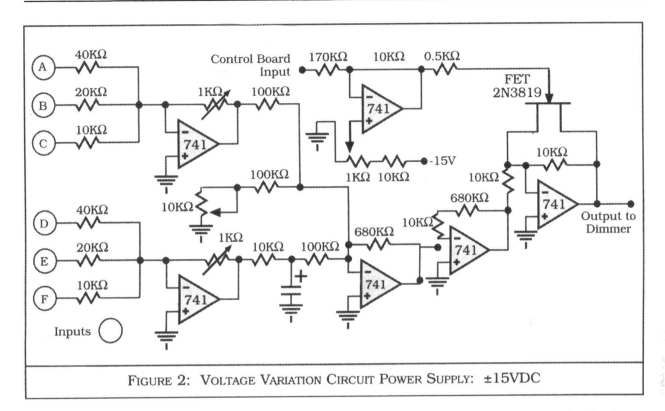

FIGURE 2: VOLTAGE VARIATION CIRCUIT POWER SUPPLY: ±15VDC

its are sent to an op-amp to be translated into voltage values that change slowly. A third voltage value with a potentiometer allows for setting the base level of brightness of the effect from which the changes occur. These three voltage values are added together proportionally in a summing amplifier to equal the voltage range required by the dimmer. The additional amplifier stages accommodate the polarity needs of the control voltage and input of the controller signal.

Through experimentation I found that the smooth changes still occurred too frequently. To remedy that problem, I arranged to send the second three digits through additional flip-flops to reduce their rate of change to one-half the time of the abrupt changes.

Built into the circuit are three variables to allow adjustment of the effect to suit individual needs and tastes. The clock pulse that controls how quickly the voltage digits change is adjustable through variable resistors from a little less than one pulse per second to several hundred pulses per second. Both the abrupt and smooth change op-amps have variable resistors to allow adjustment of the proportion of one to the other. I have already mentioned the variable base level of brightness of the effect. There is one additional potentiometer near the dimmer input to balance the input signal for the particular control voltage required by the system. This pot would normally be set once and then left alone.

Although this circuit was conceived as a television effect, it seems potentially useful for other effects. Fire comes to mind immediately. Because the circuit controls the output of an entire dimmer, the effect is limited in scope only by the capacity of the dimmer.

 za za za

Creating a lighting effect that works in rhythm with a music cue can be easily accomplished. Tinkerbelle's magical appearances would be one example. The idea is a simple, inexpensive and yet effective way of achieving this type of effect. First, decide on how you would like the light to hit the stage. Install the effect where you would normally place an instrument to light this area.

MATERIALS AND EQUIPMENT

A Loudspeaker (one that will be used solely for this effect)
Broken Mirror Pieces
A Heavy-Rubber Balloon
A Lighting Instrument (preferably one with a fairly direct and sharp beam)

PROCESS

1. Cut the balloon open so that you have a flat piece of rubber. Stretch this over the face of the speaker and attach with any reasonable adhesive. If the rubber is heavy enough, hot-melt glue will work nicely. "Super glues" also work well.

2. Using the same adhesive, glue a number of mirror pieces to the balloon covering the speaker.

3. Install the speaker in the chosen position. Patch the speaker into the theatre's sound system, giving it a separate output so that its volume can be adjusted independently.

4. Focus the instrument at the face of the speaker. Adjust the speaker and the instrument so that the mirror's reflected light hits the desired area on stage.

Passing sound through the speaker will make the balloon vibrate and the reflected light "dance" in rhythm to the sound.

NOTES

The effect will work with any type of sound going through the speaker, but to optimize the vibrations the speaker's frequency response should be suited to the frequency range of the sound being fed to it. Be especially careful to avoid setting the lighting instrument too close to the speaker: it will melt the balloon.

❦❧❦❧❦

Painting

Theatrical Applications of Aniline Dye

Theodore G. Ohl

Several types of aniline dye are available. A very useful type for theatrical applications is soluble in either water or alcohol. A wide color palette is available with aniline dyes. They are also quite inexpensive, as aniline dye powder is extremely concentrated.

To mix a concentrated dye solution dissolve about a teaspoon of dye in a quart of boiling water. This concentrated solution can be stored in closed containers until needed and then diluted with water to achieve the desired color intensity. For applications that require flame-retardant qualities, the solution can be diluted with water in which flameproofing salts have been dissolved.

Shellac-based aniline dye can be used to give a deep, translucent color to surfaces to which water-based dye will not adhere. To use shellac as a medium and binder, use a quart of unheated alcohol instead of the boiling water, and mix this concentrated solution into shellac to achieve the desired color intensity. Lightly saturated shellac-based dye can also be used to achieve a rich, toned-down surface.

A major drawback to the use of aniline dye is its tendency to bleed through subsequent coats of paint. This problem can be circumvented by applying a topcoat solution of bronzing powder and white shellac. Mix enough bronzing powder into the shellac to produce a thin opaque film on the dyed surface. The shellac should be stirred frequently while being applied to keep the bronzing powder evenly mixed. This film protects subsequent layers of finish — unless the finish contains alcohol, which will weaken or dissolve the shellac and permit the dye to bleed through. Further, a finish containing ammonia (such as high-gloss latex paint) will slightly damage a single coat of shellac and bronzing powder. Two coats of shellac and bronzing powder will usually prevent the aniline dye from bleeding through.

HEALTH AND SAFETY NOTES

Use care when working with aniline dye: it is a carcinogen. Minimal precautions should involve the use of barrier creams; non-permeable, chemical-resistant gloves; and a face mask if not a respirator.

❧❧❧❧

The following is a step-by-step process for obtaining a metallic finish on props.

BASE COAT

For the best results use an acrylic paint. On large props a mixture of clear latex and dry pigment can be used. Metal finishes such as pewter, silver, and steel can be achieved with a base coat in the blue to black range. Golds, bronzes, and brasses should be based in reds, browns, or greens, depending on the desired look.

TEXTURE

Go over the base coat with millinery lacquer and bronzing powder, which comes in a variety of different metals and should be used rather sparingly. Dip your brush into the millinery lacquer and then into a small amount of bronzing powder. Apply it to 75% to 85% of the surface area, leaving random patches of the base coat showing.

FRENCH ENAMEL VARNISH (FEV)

Give the prop a coat of French enamel varnish:

> 1 pint Denatured Alcohol
> Approximately 2 tablespoons White Shellac
> Fiebing's Dye®, (color and amount depend on the color desired). Fiebing's Dye®, a leather dye available in many colors, is a recognized health hazard. Insist that anyone who works with it wear gloves and other protective clothing as well as a respirator with an appropriate filter.

The varnish shouldn't be shiny when dry. If it is, dilute the mixture with a little more alcohol.

SHADOW

Using the base coat, shadow in about 20% of the prop's surface.

HIGHLIGHT

Finish the process by highlighting about 3% of the prop's surface with a combination of millinery lacquer and bronzing powder, a shade lighter than that used before.

The look obtained with this process can vary so much that it is important to experiment with each step in combination with the others before trying a finished piece. Each step should be dry before going on to the next. If, after a long period of storage, the finish has started to dull, it can be revitalized by starting with the FEV step and following through to the end.

SOURCE

Hunter Nesbitt Spence.

Frosting Plexiglas® Windows with Beer and Epsom Salts

Curtis Hardison

One easy way to treat Plexiglas® windows so that they appear to have been frosted by the cold is with a mixture of beer and epsom salts. This method is quick and much less laborious than sanding the Plexiglas® with sandpaper. The beer acts as a medium and as a binder for the epsom salts. After the beer has evaporated, the salts remain on the Plexiglas®, giving it a frosted appearance.

PROCESS

1. Mix the beer and epsom salts. A 2:1 mix of beer to salts seems to work well, though an increase in the amount of salts will result in a heavier frost.

2. Sponge the mixture onto the Plexiglas®. The Plexiglas® needs to be clean and laid flat to prevent the mixture from running before it dries. As the mixture dries, add more frost to smaller areas as desired.

NOTES

The mixture will take about twelve hours to dry, and there is no need to seal or treat the finished surface in any other way. After it has dried, "de-frosting" areas of the Plexiglas® can be done with a sponge and clean water. The frosting is very durable and should last the run of a show. One other advantage to this approach is that the Plexiglas® can be cleaned and reused after the show is over.

SOURCE

Sally Barr.

One of the many problems faced by technicians, painters, and designers is that of achieving believable relief textures on stage. Here is an inexpensive process that can produce textures ranging from stucco to leather and is so simple that anyone — even those without scene-painting experience — can master it in a short amount of time.

MATERIALS

The process uses two basic materials: paper products and flexible glue. The paper chosen depends on the texture to be created. Used paper towels simulate rice mats or leather, toilet tissue resembles floorplanks, and paper doilies work well for wallpaper patterns. Wrinkled wrapping tissue looks convincingly like stained glass, and wadded paper napkins resemble stucco. The possibilities for different textures with this process are almost limitless.

The flexible glue has many advantageous properties. It is easy to work with, has no toxic properties, and dries faster than most other glues. Its sheen makes it ideal for use with finish glazes. Further, since it remains flexible when cured, it wears very well as a floor surface and can even be mopped. In combination with whatever paints, dyes, or bronzing powders you use, you can create an endless number of realistic textures with these two materials.

EXAMPLE APPLICATION

Here is one idea using standard wrapping tissue to achieve high-quality wood grain textures ranging from paneling to wooden beams, logs, and bark.

1. Wrinkle the wrapping tissue up the long way, one sheet at a time. Rather than wadding it up into a ball, gather it along its full length. Prepare a number of sheets at once. Tear a ragged edge along the top and bottom of each piece so there will be no defined line where the pieces overlap.

2. Next, mix a batch of the flexible glue, cutting the glue with water. A 2:1 water to glue ratio is commonly used, but a different ratio may work better in some applications.

3. Apply the glue generously to the surface to be textured. Lay the tissue paper onto the surface with its wrinkles oriented in the desired grain direction. Without crushing the wrinkles, dab it with fingertips or a brush to keep it in place.

4. Brush a second coat of glue over the applied tissue, being careful to keep from flattening the wrinkles. The depth of wrinkles left will determine the ultimate roughness of the texture.

5. Once the piece is dry the surface is ready to paint. In painting, do not worry about leaving some small sundays: the dye glaze overcoat described below will heighten the contrast between the painted and unpainted areas, and deepen and enrich the finish.

6. While the paint is drying, mix the dye glaze. Though the proportions of components depends entirely on the desired finish, here is one possible recipe:

    > 1 gallon Water
    > 1 teaspoon Aniline Dye dissolved in water
    > 2 cups Flexible Glue

7. Apply the glaze to the painted and textured surface, taking care this time to cover every part of the surface. Avoid leaving runs and drips, which cannot be covered once dry.

SUPPLIERS

Several different types of useful paper products can be obtained in bulk and at reasonable cost from paper supply houses. It is also possible to find manufacturers who will donate paper products that are either the wrong item or size for the jobs they are doing. Be wary of paper products that come with patterns dyed into them. Such patterns can be used with stunning success for, say, a wallpaper pattern, but the dyes they use might bleed through any attempt at a paint overcoat and remain inappropriately visible in other surface finishes.

Two suppliers of flexible glues are . . .

> Swift's Products (see local Yellow Pages)
> Spectra Dynamics of Albuquerque, New Mexico.

HEALTH AND SAFETY NOTES

Use care when working with aniline dye: it is a carcinogen. Minimal precautions should involve the use of barrier creams; non-permeable, chemical-resistant gloves; and a face mask if not a respirator.

SOURCE

Don Hannon.

෫෮෫෮෫෮

A photo-mural is, roughly, any photo-enlargement larger than 20" x 24". Commercial black and white mural processing costs about $5.00 per square foot; mounting, toning, and special negative preparation are extra. It is possible to process black and white photo-murals at about a quarter of the commercial cost. The three steps are shooting and choosing the negative, testing and printing, and mounting and finishing the mural.

SHOOTING AND CHOOSING THE NEGATIVE

The negative must be clean, sharp, fine-grained, and of relatively high contrast. For shooting negatives of print material, use a very fine-grained film such as Kodak Plus-X Pan Professional®. Set the original under a clean piece of glass on a matte-black surface and shoot with a tripod-mounted camera. The lighting should be diffuse and color corrected to 3600°K. Take at least 3 shots of each print, subtly changing the f/stop and focus. Have the film processed commercially. Crop and mount the negatives in plastic slide mounts. For shooting live, using a faster speed film such as Kodak Tri-X Pan® will allow you to shoot in lower light conditions without a large increase in grain. Set the shutter speed at no slower than $\frac{1}{30}$ of a second to stop motion. If your light meter does not register at this speed, you can push the film by adjusting your camera to as much as twice the indicated speed. Indicate how much you pushed the film when you have it processed. For each negative, have an 8x10 print made to use as a guide when processing the paper.

TESTING AND PRINTING

Kodak mural paper, contrast grade 3, is the only quality paper available for printing large murals. It comes in rolls of up to 4'-6" x 98'-0" and costs approximately $0.50 per square foot. Keep the paper in its light-proof container until ready for use. Expose and process the paper in separate exposing and developing areas set up in advance to avoid destructive delays during printing. Figure 1 illustrates a good layout. Maintain darkroom conditions in both areas: use photographic safelights and keep the room temperature at 68°.

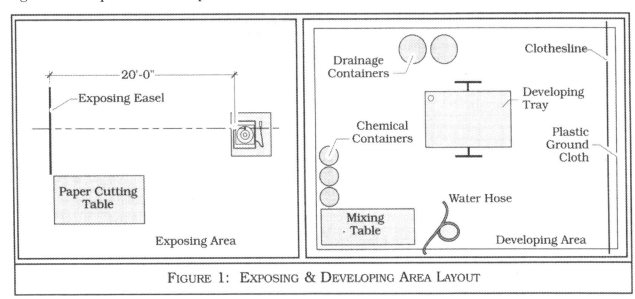

FIGURE 1: EXPOSING & DEVELOPING AREA LAYOUT

In the exposing area, set up the projector equipped with a diffusion card and a light-proof plywood cover. See Figures 2 and 3. Keep the projector's lens level with the center of the exposing easel, a framed Homasote® panel gridded into 3" squares and 12" larger than the mural in both directions.

FIGURE 2: DIFFUSION CARD & LAMP HOUSING

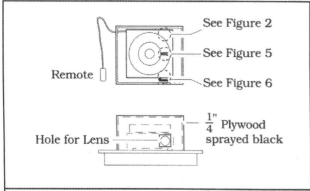

FIGURE 3: PROJECTOR & COVER

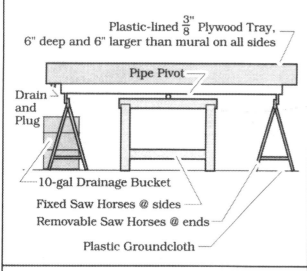

FIGURE 4: DEVELOPING TRAY

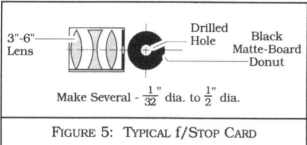

FIGURE 5: TYPICAL f/STOP CARD

In the developing area set up the developing tray as shown in Figure 4 on a large plastic ground cloth. Have ready 3 five-gallon plastic buckets containing the processing chemicals: developer; fix; and hypo cleaning agent. For best results, use Kodak Dektol® developer diluted with one part water to one part working developer solution. The temperature of all chemicals and water should be between 65° and 70°F. Slight changes in temperature of the developer can affect the density of prints. Also have ready 2 ten-gallon plastic drainage buckets, rubber gloves, water hoses with hot and cold running water, a water thermometer, and a timing clock.

Before printing the mural, run tests to determine the best combination of exposure and development time. Choose the print with the sharpest focus and best contrast range. Remove the projector's light-proof cover, put the corresponding negative in the projector, and focus the image on the exposing easel. Place the $\frac{1}{16}$"-f/stop card shown in Figure 5 in front of the lens and adjust the focus. With a footcandle meter check the light level at the easel — it should be $\frac{1}{4}$fc with the negative, 2fc without the negative. With the image focused on the easel, pick a 6" square that is of medium density and mark it with a push-pin. Reverse the slide tray to a blackout slide shown in Figure 6 and lower the light-proof cover over the projector.

Cut a 6" square test piece of mural paper. Lay a Kodak Projection Print Scale on the piece of mural paper and push-pin the test piece and scale to the easel over the square you picked earlier. Turn the projector on and forward the tray to the negative. Expose the test for one minute, then reverse the tray to the blackout slide.

Process the test piece in the premixed chemicals. Wear rubber gloves and agitate the test in each container as follows: water rinse, 10 seconds; developer, 3 minutes; water rinse, 15 seconds; fix, 5 minutes; water rinse, 1 minute; hypo, 2 minutes; and water rinse, 5 minutes.

On the test print, you should see a circle divided into wedges that get progressively darker. These relate to the divisions on the print scale. If the print is very dark, use a smaller diameter f/stop card; if it's washed out, use a larger diameter one. Pick the wedge that is closest in density to that of the 8x10 print. The proper exposure time is marked in the corner of the wedge. Turn on the projector and pick three 6" squares; one dark, one light, and one with a lot of detail. Pin up 3 pieces of mural paper to these areas and expose them for the timing you chose. Process these as before and check their density against the 8x10 print. If the light test area is washed out, you can "burn in" detail by exposing the print for the allotted time and then exposing the light areas for a few seconds more by waving a burning tool about 1' in front of the lens to block out everything but the light area. If the dark areas are too dark, you can "dodge" these areas by waving a dodging tool about 1' in front of the lens, blocking light to the darkest areas during the normal exposure time. Figure 7 shows both tools. If contrast is low, try a combination of underexposing and overdeveloping.

Develop the mural as follows: rinse with water for 10 seconds; drain the water into one of the ten-gallon containers; plug the drain and pour in the developer; remove the end sawhorses and seesaw (rock) the tray for 3 minutes; check the highlight area for detail. Under safelights, prints will appear deceptively dark; if necessary, "burn in" highlight detail with a sponge soaked in fresh developer; drain the developer back into its container and rinse with water for 15 seconds; plug the drain and pour in the hypo; rock for 2 minutes; drain and wash with water for 5 minutes; hang the print up to dry or wet mount it immediately.

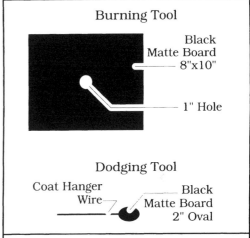

FIGURE 6: BLACKOUT SLIDE

FIGURE 7: EXPOSURE CONTROL TOOLS

MOUNTING AND FINISHING THE MURAL

The mounting surface must be rigid and smooth, like lauan or Masonite®. The mounting adhesive can be any commercial chemical mounting adhesive. More economical is wet mounting with either printers' padding, or a 1:1 solution of white glue and water applied evenly to the mounting surface. The wet mural can then be pressed in place with a damp sponge. Wrinkles can be removed by placing a wet rag in the center of the mural and allowing the edges to dry first.

Aniline dyes and magic markers are best for coloring prints, but commercial toners such as Kodak sepia toner must be used for complete retoning. Lacquer seals prints best because it will not yellow. Clear latex is also satisfactory, but it will not stand up to outdoor use. Shellac and FEV can be used for aging.

SPECIAL PROBLEMS AND WARNING

Under intense stage lights fading may occur. Conditions of humidity below 20% may cause brittleness, and above 50% may cause discoloring and buckling. Recent studies have shown that photographic chemicals are highly toxic and carcinogenic. Wear rubber gloves and a respirator.

The process described here is an inexpensive method of quickly duplicating the look of a high-contrast black and white photograph in a large scale. It is significantly cheaper than a painting produced by machine from a rendering, and it is quicker than painting a photo blow-up from a projected image. Three people painted about 1200 square feet of flattage in five days for a Yale School of Drama production.

The method is based on the "frisket principle," whereby objects are laid on a surface before it is spray painted. When the objects are later removed, the shapes of the objects remain unpainted. The process lends a certain freedom to the painting process but requires some experimentation with materials.

In this application, a sequence of painting steps results in a photographic image comprising layered shades of gray.

PROCEDURE

If the fabric used is to be mounted on a flat frame, paint it before framing to avoid inaccuracies that would otherwise occur where the fabric sagged or paint sprayed under the frisket.

1. Staple the fabric to be painted onto the floor and prime it with white paint.

2. Mark important reference points on the fabric. Include the edges of the flat or drop, and draw in other significant elements of the photo.

3. Assemble the frisket: place wood, rope, cloth, paper, and other materials on all of the areas that are to remain white or a shade of gray. These frisket objects should, when possible, be the same material as the objects in the photograph, *e.g.*, use rope for clothesline and cloth for laundry.

4. Use gravel and sawdust for texture in areas that are to look soft and broken up.

5. Spray a mist of good quality black paint around the edges of the frisket materials. The object is to produce a shade of light grey — not black.

6. Remove the materials that represent the next-darkest objects in the photograph, and spray another layer. Continue alternately removing objects and spraying paint until all of the frisket materials have been removed. The areas uncovered last remain the whitest.

7. To create large-scale textures like brick or stone, make a hardware-cloth screen in an easily handled size. Glue paper cutouts of appropriate shapes to the screen. Apply joint compound to create some holes in the screen and polyurethane to waterproof and stiffen the paper. Paint sprayed through the screen will reproduce the shapes of the bricks or stone with a "dotty" look.

8. Glaze the entire painting with warm and cool washes of watered-down casein tinted with dyes to match the photograph.

9. With a Hudson-type sprayer, spray the finished painting with a 3:1 mixture of clear flat and clear gloss latex. This finish coat both seals the painting and gives it the matte look of a photo print.

NOTES

This process is used to duplicate an existing photograph. Study the photo to find appropriate textures and materials.

Avoid being too specific about a texture. A brick mask, for example, should be composed of shapes that merely suggest the shape and repetitiveness of brick rather than neatly arranged rectangles.

Always spray through or around a frisket. Arbitrary spraying without a frisket defeats the advantages of the process and can spoil the results.

Mark the completed painting for identification and orientation to avoid confusion during framing.

SOURCE

Lisa Frank.

Photocopies of drawings or photographs can be reproduced on any type of natural-fiber fabric. The simple process outlined in this article applies to both color and black and white photocopies and assures exact duplication of any pattern. This process is useful in making your own fabric patterns for limited size applications such as upholstery. It reproduces multicolored patterns faster and more accurately than stencils. Reproductions of color or black and white photographs on pillow covers, clothing, towels, and other small items can be accurately and easily accomplished with this process. Photocopying (about 40¢ for color, 5¢ for black and white $8\frac{1}{2}$" x 11") is relatively inexpensive compared to the equivalent cost in time involved in painting or stenciling.

FABRIC

Any fabric made from a natural fiber such as cotton, flax (linen), wool, or silk can be used, though cottons and linen work best. In general, the tighter the weave of the fabric, the sharper the image on the fabric will be, and conversely, the looser the fabric weave, the fuzzier the image. Thus, cotton-backed satin would be a good choice for reproducing hard-edged patterns and designs; and light-weight muslin, a good choice for reproducing those with a softer quality. Lettering on photocopies will appear in reverse on the fabric. This problem can be solved, however, by photocopying a reversed transparency (slide) of the original. Note that photocopy ink will bleed when the fabric is washed unless the ink has been set by ironing after completing the transfer process.

MATERIALS

Any absorbent Material	A Rag
Photocopies	Fabric
Acetone	White, Iron-On Pellon (optional)

PROCESS

1. Obtain enough photocopies to duplicate the pattern as many times as desired. Each reproduction requires its own photocopy.

2. Back the fabric with pellon if it needs to be reinforced.

3. Place the fabric, pellon-side down, on the absorbent material. Place the photocopy, face down, in the desired position on the fabric.

4. Hold the photocopy in place and firmly rub the back with a rag soaked in acetone until the photocopy is saturated.

5. Peel the photocopy carefully from the fabric and allow it to air dry thoroughly before use.

SAFETY PRECAUTIONS

Acetone is extremely flammable, and should not be used near an open flame. Acetone is also very toxic; use in a well-ventilated area and wear gloves and a respirator throughout the process.

❧❧❧❧❧

To translate a small rendering or painting onto a large set element, the scenic artist may first make a full-scale line drawing, often called a "cartoon." In order to achieve an accurate drawing, a set painter makes the cartoon from a full-scale projection of the original artwork. Any opaque, carousel, or large-format projector can reproduce the original on a wall or a screen perpendicular to the floor. But for those artists who prefer to paint their material on the floor, either the Flying Periscope or the Cart Periscope described below would allow the cartooning to be done in the same location as the actual painting.

CARTOONING WITH A FLYING PERISCOPE

The Flying Periscope is for use by painters who work on a stage floor with a fly system overhead. The size of the image projected onto the floor is adjusted by the height to which the projector is raised in the fly loft.

Fly the periscope to a reachable height, focus the projector, and adjust the tilt of the mirror until the projected image on the floor does not appear distorted. Allowing the remote controller to hang down, raise the batten until the projection is near the desired scale, and re-focus. Fine tune the height and focus until the projected image matches the full-scale dimensions you have laid out on the floor.

CONSTRUCTING THE FLYING PERISCOPE

Piece **A** in Figure 1 is a hollywood flat covered with mirrored Mylar® or any other material with a specular surface. The flat is attached to the two extended side pieces of projector platform **B** at the pivot points **D** so the angle of tilt may be adjusted. The mirror must be far enough away from the platform that the light beam does not burn the mirror surface and is not obstructed when reflected downward. This distance, as well as the size of the projector platform, depends on the type and size of projector used, the heat output of its lamp, and the dimensions of its beam. Frame **C** attaches to a batten with U-bolts or C-clamps. These guidelines are, of course, far from a definitive construction approach.

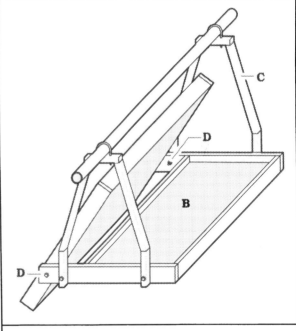

FIGURE 1: THE FLYING PERISCOPE

CARTOONING WITH A CART PERISCOPE

Adjusting the size of the image with the Cart Periscope is not as easy as with the Flying Periscope, but if a fly stage is not available the Cart is a simple alternative. The scenic artist first grids the original work into workable squares. He or she then shoots a slide of each grid square that includes some portion of any adjacent squares. In cartooning, start at one corner of the drop and adjust the angle of the mirror and the focus of the projector until the image is not distorted. Tailor the size of the projected image to the grid square by raising or lowering the projector platform. When the corner square is cartooned, move to an adjacent square, change the slide, fine tune the focus and alignment, and then continue cartooning. Move from one square to its neighbor until the drawing is complete.

CONSTRUCTING THE CART PERISCOPE

The mirrored frame **A** and the projector platform **B** in Figure 2 are constructed the same as they were for the Flying Periscope. In order to add stability to the cart, it is necessary to place extra weight on the counterweight platform **C**. Attachment holes **D** in the framing members connecting platforms **B** and **C** enable the projector to be raised and lowered for size adjustment.

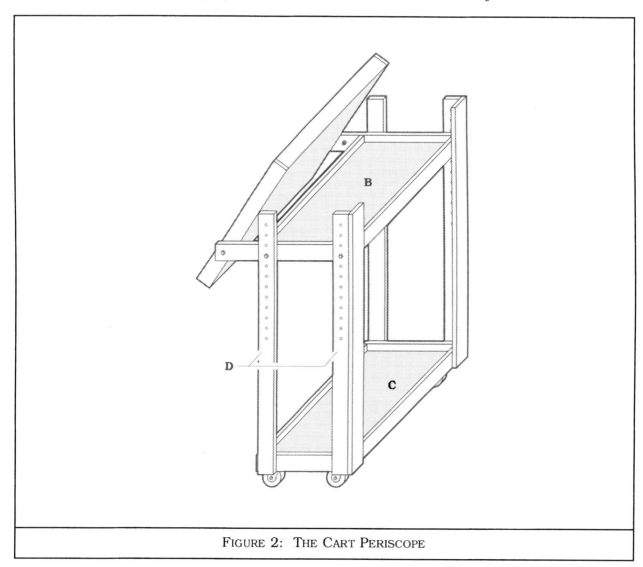

FIGURE 2: THE CART PERISCOPE

THE IMAGE DISTORTION PROBLEM

Projecting at such a large scale may cause the image to curve at the edges or fall out of focus in the corners. The use of better-quality lenses and high-powered projectors lessens the effects of distortion. Wide-angle lenses, which spread the image over a large area in a relatively short throw, are ideal for the Cart Periscope. The fish-eye effect is not nearly as pronounced at the projected scale as it is in those familiar photographs of leering faces. Finally, leaving a part of the neighboring squares in the gridded rendering when shooting slides for the Cart Periscope insures that the grid to be cartooned will be distortion-free even if the border itself is not.

FINAL NOTE

When detail and accuracy are demanded on a large-scale reproduction, say of a large version of a well-known artwork, projection is generally the solution. With either the Flying or Cart Periscope a scenic artist can accurately cartoon in almost any shop or on any stage. You may encounter problems in the initial setup but if you keep in mind that it is distortion that inhibits accuracy, you should end up with very satisfying results.

❧❧❧

For years I have been stenciling wallpaper patterns and have tried a variety of published techniques. Each method involved some drawback. Ordinary stencil paper will not last long enough to complete a large project even if shellac or some other sealant is applied, so multiple masters need to be cut. If you must work on a vertical surface and the pattern is lacy, the weight of the paint often pulls some parts of the stencil away from the wall, and delicate sections eventually break off.

In searching for a more suitable and durable material, I cut a small pattern out of a plastic "Garage Sale" sign available at any hardware or discount store. It was extremely easy to cut with an X-Acto® blade, and even lacy sections did not break away after multiple uses. The original sign was too small to be practical for many stenciling needs, but I observed that many in-store signs are made of this material to promote products and advertise specials. At the time, I did not know that the material is 15-mil polystyrene, available in sheets of several sizes from plastics suppliers. I did learn, however, that even some of the larger signs can be obtained free of charge after a short discussion with the store manager, since stores dispose of them anyway after the sale or special has ended.

Durability is only one of polystyrene's advantages as a stencil material. It is very lightweight, and large, stiff stencils can be handled easily by one person. If your stencil has small sections that are difficult to keep in contact with the surface, you can spray the back side of the stencil with a light coat of Scotch 77® adhesive and avoid underspray by temporarily gluing the stencil to the surface. After a light dusting with the adhesive, the stencil will stick to the wall by itself with no underspray or drips at all. You can remove and reapply the stencil four or five times before you need to apply another coat of the Scotch 77®. When the stencil is removed from the wall, the adhesive does not leave any residue. The front surface of the stencil can be cleaned easily with a wet sponge after each application to prevent paint buildup. When you are finished, your stencil can be saved for future use. If paint does dry on the stencil, simply place it in a pan of water for a day or so and the paint will separate easily from the plastic. Scotch Natural Cleaner® can be used to remove residues of the Scotch 77® adhesive, and you'll have a like-new stencil.

¿▲¿▲¿▲

TECHNICAL BRIEF

Props

Snowballs for the Stage	*Theodore G. Ohl*
Artificial Canapes	*Sharon Braunstein*
Stage Blood	*Randy Fullerton*
A Remote-Controlled Portable Water Source	*Cosmo Catalano, Jr.*
Circuitry for a Remotely Dimmable Portable Lighting Practical	*Donald R. Youngberg*
Remote-Control Live Fire	*Chris P. Jaehnig*
High-Volume, Low-Cost Modeling Clay	*Jon Lagerquist*
Gelatin Molds	*Mark Shanda*
A Remote-Controlled Flash Effect	*Steven A. Balk*
Growing Flowers Onstage	*Scott Servheen*
Break-Away Glass: Formula and Process	*Bill Ellis*
A Butane Torch for Use Onstage	*Rod Hickey*
A Safe Lamp-Oil Torch	*Alan Hendrickson*
A Light-Sensitive Portable Practical	*Tim Fricker*
Faking Waterproof Beer Can Labels	*Chris Higgins*
Quick Casts in Under Two Hours	*Christopher Sibilia*
Liquefied Auto Body Putty	*Dr. Ronald Naverson*
A Very Useful Snow Machine	*Richard Gold*

INGREDIENTS

> One Personal-size Bar of Ivory® Soap
> One Cup of Cold Water

EQUIPMENT

> Cheese Grater
> Blender
> Electric Mixer
> Ice Cream Scoop

Coarsely grate the soap. Place the cold water, then the grated soap, into the blender, and immediately liquefy the mixture for 2 to 4 seconds or until it is foamy. Avoid blending for too long a time (that is, until the soap is completely dissolved), as the soap mixture must still have a bit of texture to whip properly. Snowballs that have been blended too well will be rubbery and not break convincingly on impact. Pour the blended soap into a bowl and beat on high speed until the mixture "peaks," having a meringue-like consistency. Using the ice cream scoop, dish the whipped soap from the bowl onto a piece of waxed paper. One batch of soap should yield 5 or 6 large snowballs.

Let the snowballs dry overnight, then turn them and let them dry overnight again. The dry snowballs may then be lightly scored into quarter sections to facilitate breaking. Finished snowballs are usable for 4 to 5 days, but after that they become dry and will not break properly.

SOURCE

Hunter Nesbitt Spence.

Artificial canapes can be a valuable alternative to real ones, since food can spoil and replacing it is costly. These canapes can be used alone or as supplements to real canapes.

BREADS AND CRACKERS

BROWN AND RYE BREADS

Choose a yellow cellulose sponge. Yellow is the ideal choice because it is the most neutral-color sponge available and will take paint best. If you cannot get a sponge the thickness of the bread you would like, you can slice a thicker sponge with a razor or band saw. Cut the sponge to fit the shape you want for your canapes. To color, mix scene paint to the appropriate color and soak the sponge in it. It may be necessary to dip the sponge several times to achieve the right color.

WHITE BREAD

Cut white Styrofoam® to the desired shape. Coat with tissue paper and diluted white glue.

CRACKERS

Cut shapes out of $\frac{1}{4}$" foam-core board. Compress to an irregular thickness of about $\frac{1}{8}$" by using your knuckles. For additional depth and shaping, use a blunt object like a screwdriver handle. The final texture and color can be achieved by stippling the crackers with scenic dope applied with a small irregular piece of sponge. If you are creating a light-colored cracker, you can use the natural dope color. If the cracker is to be another color, add dry pigment to the dope.

TOPPINGS

SOFT CHEESE AND SPREADS

Tint powdered wood putty, such as Durham's Rock Hard Wood Putty®, with acrylic paints to create the desired color. If the topping is to be applied decoratively, you can use a standard pastry tube available at any kitchen supply store. For grainier foods, such as ham spread, add sawdust to the putty.

PIMENTOS AND OLIVES

For pimentos, dip red fabric in shellac and drape across the canape. Beads can represent green or black olives, whole or sliced.

CAVIAR

Shellac small black beads together.

GARNISHES

Decoration adds a final touch of realism. Plastic vegetables, such as carrots and peppers, can be chopped, diced, or grated and used on spreads or cheese. Plastic greenery, particularly parsley fern, is also good for topping off a piece.

SOURCE

Hunter Nesbitt Spence.

In the following stage blood recipes, specific quantities of the ingredients can be determined only through experimentation. There are several factors to consider: the quantity and texture of the blood, the surface to which the blood will be applied, and the light under which the blood will be seen.

INEXPENSIVE, EASY METHOD

    KY® Jelly
    Water
    Food Coloring (red and green)

Combine red food coloring with KY® Jelly. Dilute this mixture to the desired consistency with water. Add the green food coloring as a toner. Mix thoroughly. This mixture dries quickly and looks chalky when completely dry. It is not good for long scenes and also stains fabric and skin.

BLOOD IN THE MOUTH

    White Corn Syrup
    Corn Starch
    Water
    Food Coloring (red and green)

Combine white corn syrup with a small amount of corn starch for body. Dilute this mixture with water until the desired consistency is obtained. Add red food coloring and then slowly stir in green food coloring until the desired color is achieved. This mixture is edible and easy to make, but it is sticky and does stain fabric and skin.

    Carnation Instant Breakfast® (chocolate or strawberry)
    Ateco® Food Coloring (red and green)
    Water

Combine all ingredients until the desired color and consistency is achieved. Ateco® is a highly concentrated paste food coloring available at bake shops or kitchen supply stores. "Christmas Red" and "Leaf Green" were used in this recipe. As a result of the food coloring, this mixture also stains fabric and skin.

COMMERCIALLY PREPARED BLOOD

    Reel Blood®
    Jack Sperling Studio
    13639 Vanowen Drive
    Van Nuys, CA 91405
    (213) 780-7715

A very good product, one gallon costs $40.00 but lasts a long time. The mixture hardly stains, can be diluted, and can be used orally — it tastes like cinnamon. It has a sugar base and becomes sticky. This product dries normally and does not appear chalky.

OTHER COMMERCIAL SOURCES

> Rosco Laboratories
> 36 Bush Avenue
> Port Chester, NY 10573
> (914) 937-1300
> $5.50 for 6 ounces
>
> Special Effects
> 18 Euclid Avenue
> Yonkers, NY 10705
> (914) 965-5625
>
> Theatrical Services and Supplies
> 2050 Brook Avenue
> Deer Park, NY 11729
> (516) 242-5454
> $5.50 for 8 ounces (External use only)

GENERAL COMMENTS

Most blood mixtures can be combined with shredded cheesecloth to resemble ripped flesh. The cheesecloth can be attached to the flesh with liquid latex. When the latex has dried, a coating of blood is applied.

A convenient way of dispensing blood within the actor's mouth is by using commercially purchased gelatin capsules. These are obtained from pharmacies and can be filled by using an eyedropper. Once they are filled with blood, the capsules last about 30 minutes — longer if refrigerated. Prolonged refrigeration stiffens the blood mixture, however, and time should be allowed for the capsule to soften at room temperature.

Another way of dispensing the blood is by using a small bloodbag — a corner of a cheap plastic sandwich bag tied off with a string or rubber band is sufficient. This method provides a convenient way of presetting blood on the set or in an actor's pocket and is extremely effective when bitten into and spit out by the actor.

Large quantities of blood can be dispensed by filling a sponge with blood and placing it in a plastic bag on the set or in a plastic-lined pocket on the actor. Most stage bloods stain both fabric and skin. A sample of any blood used should be tested by the costume department. Cold water is effective for removing blood stains on fabric, and bleach can be used as a cleansing agent. The fabric should be soaked in cold water before the stain has dried.

SOURCE

Hunter Nesbitt Spence.

❦❦❦❦

A Remote-Controlled Portable Water Source

Cosmo Catalano, Jr.

In a production at Penn State University we needed a working hand pump on stage. For various reasons, we could not hook up a hose to a sink or use a recirculating pump to supply the water. Furthermore, as the pump was used *ad lib* by performers during the show, we wanted to control the flow of water from the stage manager's desk in the light booth. The pump we found could not draw water on its own.

We were, therefore, presented with two problems — the supply source for the water, and the remote control of the flow. We chose not to use a Hudson-type sprayer for the supply source because we were concerned with the potability of water from a device not designed for that use. We decided instead to use a hydro-pneumatic pressure tank. These tanks are often used in houses where the water is supplied by a pump, as they provide a constant pressure without the necessity of running the pump continuously. A hydro-pneumatic pressure tank has two compartments separated by a plastic diaphragm. On one side of the diaphragm the air is pressurized to around 20 pounds per square inch. The other side is then filled with water, further compressing the air behind the diaphragm. The compressed air maintains pressure on the water, assuring an even, pressurized flow from the tank. Because the tank provides its own pressure, it can be easily used on remote or moving pieces of scenery.

To solve the control problem we used a discarded dishwasher solenoid valve that was responsive enough to simulate the pulsing of hand-pumped water. Because of the high control voltage required (120VAC), we electrically grounded the entire system to reduce the possibility of electrical shock. A ground-fault-interruptor would further reduce this possibility and is recommended.

The hand pump used on stage had no plunger or valves. A rubber hose was run through the pump, ending just short of the mouth. The other end was hose-clamped to the output side of the solenoid. After the initial actuation this hose stayed full of water, and the lag time between the actuation of the solenoid and the arrival of the water at the mouth of the pump was negligible. The input side of the solenoid was connected by a short length of garden hose to a hand valve on the hydro-pneumatic tank. Because the solenoid — an on/off device only — was not able to control the volume of the water through the system, this valve was used to preset that volume. By installing a tee fitting and another valve at the tank output, the tank can be filled without removing it from the set piece. See Figure 1.

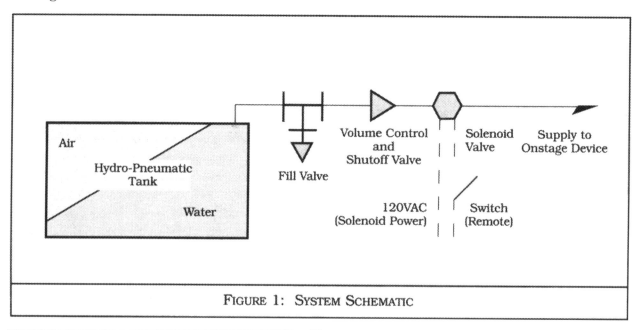

FIGURE 1: SYSTEM SCHEMATIC

The non-adjustability of volume is the major drawback with this system. If cues called for varying amounts of water, a different and more expensive solenoid valve would be needed. Another drawback is the limited quantity of water available from the tank. Our tank held about 4 gallons. Larger tanks are available at increased cost, weight, and size.

In a situation where remote control of water flow is not necessary, the output of the tank can be connected directly to standard plumbing fixtures and controlled by actors onstage.

SUMMARY OF COSTS

Hydro-pneumatic tank, 7.3 gal.	$75.00
Solenoid from old dishwasher	$0.00
50' garden hose	$5.95
Hand valve for tank output	$4.89
Hose-clamps and miscellaneous fittings	$5.00
	$90.84

The initial outlay for this system is high, but all the parts are reusable and greatly simplify the installation of any limited-use water supply.

❧❧❧❧

Circuitry for a Remotely Dimmable Portable Lighting Practical

Donald R. Youngberg

This article describes an electronic circuit that allows a battery-powered practical to be carried onto the stage, set down, and remotely switched to an external power source. Advantages of this circuit are the dimming and extinguishing of an unattended, battery-powered practical during or at the end of a scene.

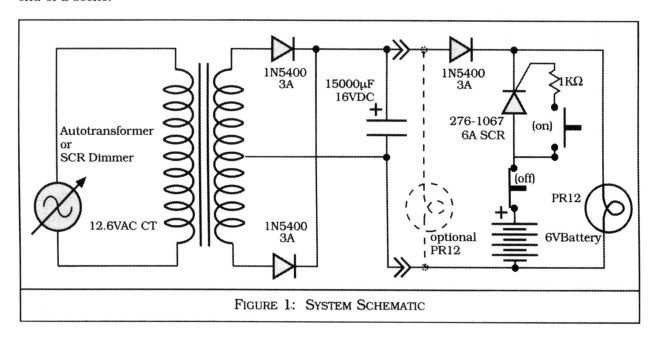

FIGURE 1: SYSTEM SCHEMATIC

The Silicon-Controlled Rectifier (SCR) in the circuit allows a lamp's power source to be switched from an internal battery to an external source. The resistor and "on" switch supply and limit the threshold gate current that triggers the SCR and turns the lamp on. When the voltage from the external source rises to a value equal to the voltage across the lamp, the current through the SCR goes to zero, turning it off. This opens the battery circuit and turns the control of the lamp over to the external power source. The "off" switch provides an easy method of interrupting the battery circuit if it were accidentally turned on.

Contacts connect the battery-powered practical to its remote external power supply. Large, thin copper plates cemented to the bottom of the practical, and spring-loaded contacts concealed in a scenic unit make this connection simple and relatively foolproof. With practice, an actor can align the baseplate rings and contacts easily.

A variable DC power supply or a rectified and filtered variable AC source can work as the external power source. Connection to an existing dimmer includes the practical in lighting presets. The optional lamp provides a way to increase the intensity of the practical after connection to the external power source. The schematic shows a possible rectifying and filtering circuit for the output of an SCR dimmer. The capacitor must be large enough to keep the lamp from flickering when the external source voltage approaches the battery voltage. The circuit described above costs about $40.00 and can be easily built by those who have some background in electronics.

SOURCE

Alan Hendrickson.

Described below is a safe and proven method of controlling live fire onstage from a remote position. The reader is cautioned that the use of any live fire effects must gain the prior approval of the local Fire Marshal. Because of the relative expense of building this system, plans for the effect should be presented and approved before construction is undertaken.

ELEMENTS OF THE SYSTEM

FUEL

The fuel most commonly used is propane. This fuel is used in most motor homes and farm buildings, and suppliers are listed in the Yellow Pages under "Petroleum—Liquefied." When handled with care it is a safe fuel. Route the fuel to the other system components with rubber hose similar to the hose used on oxyacetylene rigs. The use of rubber hose precludes establishing any unwanted electrical circuits — which might include the fuel tank.

CONTROL

The fuel is routed from the tank to a normal-closed solenoid on/off valve. From the solenoid the propane moves to an adjustable flow control valve that adjusts the height of the flame.

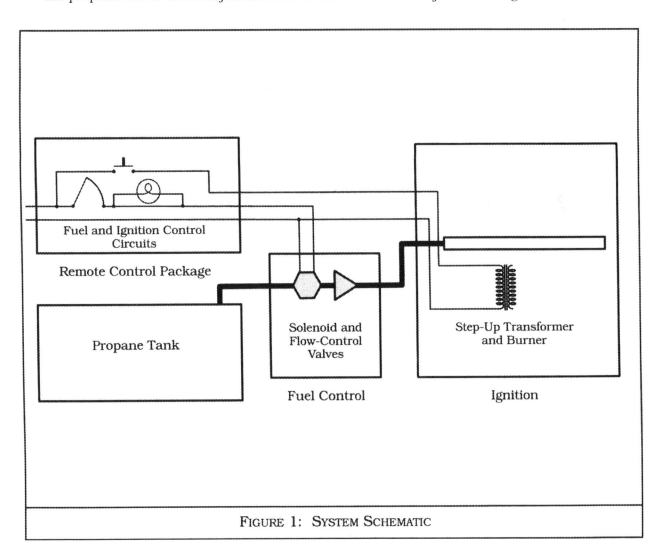

FIGURE 1: SYSTEM SCHEMATIC

BURNERS

The fuel travels to a burner that is fabricated from a length of $\frac{1}{4}$" stainless steel brake line hose bent into an appropriate shape. Holes $\frac{1}{32}$" in diameter are drilled on 1" centers to allow fuel to escape and mix with air.

IGNITION

The fuel is ignited by a spark produced by a high-voltage step-up transformer such as that used in neon light applications. Heavy-gauge, solid wire terminal leads from the transformer are bent to form an arc gap approximately $\frac{1}{4}$" above the burner.

REMOTE-CONTROL PACKAGES

This package should contain a toggle switch and indicator light for the solenoid on/off valve, and a momentary contact switch for the transformer.

SAFETY PRECAUTIONS

Observe the following before you test the system:

1. Check all connections; there must be no leaks.

2. Isolate the transformer and the terminal leads from all other parts of the system.

3. Check that all the scenic elements near the fire are thoroughly flame-proofed.

ACTIVATING THE SYSTEM

1. Open the valve of the fuel tank.

2. Activate the solenoid on/off valve, allowing fuel to enter the burner.

3. Depress the momentary contact switch controlling the transformer until the fuel ignites.

4. Adjust the height of the flame with the flow-control valve.

FINAL NOTE

A campfire using this design was built for the Seattle Repertory Theatre's production of *Savages*. To allow the flame to grow on cue, a second control line and burner were added. When the second solenoid was activated, the flame spread to the second burner. When both solenoids were deactivated at the end of the scene, the fire was immediately extinguished.

&a&a&a

The need for large amounts of modeling clay in several productions at Cal State Long Beach spurred research into low-cost clays. The local art school had an easy-to-make clay recipe using easily found ingredients.

MATERIALS

These ingredients will yield one 150-pound batch:

     3 slabs Microcrystalline Wax (a slab weighs 11 pounds)
     10 quarts of new Non-detergent 30-weight Motor Oil
     100 pounds of Talc (hydrated magnesium silicate)

DIRECTIONS

Using a large, clean tub, melt the wax into the oil over a medium flame. Add the talc slowly while stirring and heating to keep the mixture fluid. When ingredients are fully mixed, pour the clay into forms or onto a plastic sheet on a cement floor and let cool at room temperature.

FINAL NOTES

Varying the ratio of ingredients will affect the feel of the clay. The wax and talc can by found at art or pottery shops and the motor oil at auto supply houses. It may be possible to dye the clay before adding the talc, but the dye's setting ability should be checked in a small batch before mass production.

SOURCE

Herbert Camburn and Richard Johnson.

❧❧❧

Gelatin Molds

Mark Shanda

The following is an easy method for creating a flexible negative mold for use in making plaster or polyester resin castings.

MATERIALS

Knox® Unflavored Gelatin
Ethylene Glycol (permanent antifreeze)
Water

EQUIPMENT

Hot Plate
Double Boiler
Plastic or Non-porous Container (for mold making)
Coffee Can with Lid (for storage)
Positive of Desired Casting. The best results are achieved with non-porous objects such as metals, stone, glazed ceramic, glass, plastics, and Plasticine®.

In the top of a double boiler add equal parts by weight: dry gelatin mix, cold water, and ethylene glycol. Stir the mixture into a smooth slurry and begin to heat. Keep mixing over the heat until the gelatin is all dissolved and you have a semi-transparent and bubble-free viscous liquid. The mixture is now ready to be used immediately or to be stored in the coffee can for later use. The stored mixture may be made ready for mold making by remelting in a double boiler.

When making a negative mold, place the positive you wish to copy into a shallow plastic or non-porous container that is at least $\frac{1}{2}$" larger and higher than the item to be duplicated. Next pour the melted mold-making mixture over the positive to a height at least $\frac{1}{2}$" higher than the positive. Let cool slowly. You should have no air bubbles in the mold. When cooled to room temperature, remove the gelatin mold from the positive. This should be done with some care for the mold will tear, but the gelatin's moisture content will not allow it to stick to the positive. A plaster or polyester resin casting may now be made from the gelatin mold. When no more castings need to be made, the gelatin may be remelted and used for creating another gelatin mold.

Over time the gelatin mold will slowly lose water and shrink, but its life can be prolonged by wrapping it in a plastic film like Saran Wrap® or by storing the mold in an airtight container.

ONE NOTE OF CAUTION

When casting with either plaster or polyester resin to a thickness of over $1\frac{1}{2}$", sufficient heat may build up to melt the gelatin negative. Our best results were achieved with castings that were no thicker than $1\frac{1}{2}$".

SOURCES

Dennis Dorn.
Plastics for Arts and Craftsmen, Harry G. Hollander.

In a Yale School of Drama production we used an infrared transmitter and receiver to control the operation of a four to six-foot flash effect that emanated from a ceramic bowl. See Figure 1. Stage action prohibited the use of contacts hidden in the stage floor. As a safety precaution, the system had to be designed to be triggered reliably and by one device only. Radio and ultrasonic controls were ruled out because of the risk of stray radio waves or sounds that could trigger the flash prematurely.

The main design requirements of the flash effect were that it:

1. be produced by a mechanism that could be concealed inside a 1'-tall ceramic bowl having a top diameter of 15" and a base diameter of 6";

2. be capable of being carried onstage by an actor in the course of the play's action;

3. be capable of sitting unattended onstage for 15 minutes before ignition;

4. be capable of producing a single flash 4' to 6' feet tall.

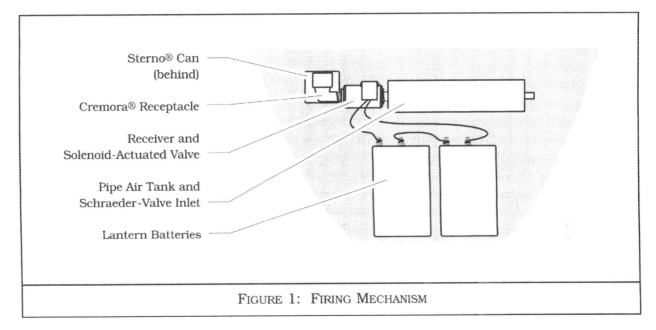

FIGURE 1: FIRING MECHANISM

THE SYSTEM PRINCIPLE

An infrared transmitter, positioned on a pipe thirty feet directly above the spiked placement of the bowl, emits a beam of light "visible" to the receiver only. When the transmitter is activated, the receiver responds by closing a circuit that operates a valve, releasing pressurized air through a container of Cremora®, the common coffee whitener. The Cremora® is blown into a suspension in the air above the bowl, and as it passes a pilot light of Sterno®, it ignites quickly, simulating a dust explosion. The system comprises two fundamental components: the remote control device and the firing mechanism.

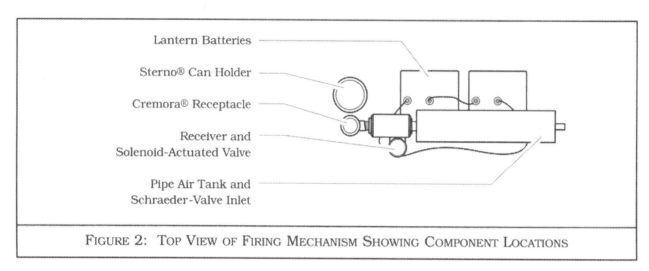

FIGURE 2: TOP VIEW OF FIRING MECHANISM SHOWING COMPONENT LOCATIONS

REMOTE CONTROL DEVICE

A frequency-paired transmitter (Banner, Inc. #SMA91E) and receiver (Banner, Inc. #SMW95R), positioned as described above, act as the remote control device. The transmitter is powered by a non-dim; the receiver, by two 6V lantern batteries connected in series. The receiver contains a built-in relay that, in this application, opens the firing mechanism's solenoid-actuated valve.

THE FIRING MECHANISM

The small, shop-built tank pictured in Figure 2 consists of a 1"-diameter schedule 40 black pipe, 6" long. The pipe is sealed with $\frac{1}{8}$" flat stock brazed across its open ends. The pieces of flat stock are drilled and tapped to accept a Schrader tank valve (#1468E-6) and a solenoid-actuated valve. Through this valve, the air tank can be charged to 80psi. The two-way, normally closed, spring-return, solenoid-actuated valve, powered by the same batteries that energize the receiver, controls the flow of air out of the tank. The output side of this valve is fitted with an $\frac{1}{8}$" to $\frac{1}{2}$" NPT reducer, drilled out to form a smooth funnel-like receptacle in which the Cremora® will sit. A sheet metal Sterno® can holder is attached to this funnel with a small hose clamp.

THE FLASH

Just before the bowl is to be brought onstage, the Cremora® receptacle is filled and the Sterno® is lit. After the bowl has been placed on spike, the device is ready to be triggered. The light from the transmitter activates the receiver's relay, energizing the solenoid-controlled valve and throwing the Cremora® into the air like a cloud of dust. As the cloud floats down, its bottom edge makes contact with the lit Sterno®, and the whole Cremora® cloud ignites in a spectacular, billowing flame.

INFRARED EQUIPMENT SOURCE

Banner Engineering Corporation, Inc.
9714 10th Avenue North
Minneapolis, MN 55441
(612) 554-3213

Recently, I was faced with the problem of devising a way to make flowers seem to grow on stage between acts. The number of flowers needed, and the time frame in which they needed to appear, prohibited a stagehand from placing the flowers onstage on cue. Thus, the growth of the flowers had to be both quiet and reasonably fast. I developed a mechanism in which each of several flowers is attached to a piston-like disc and seems to grow as its piston is pulled up a cylinder of PVC pipe and through the surface of the deck. Below is a diagram and a short description of how you might approach a similar project.

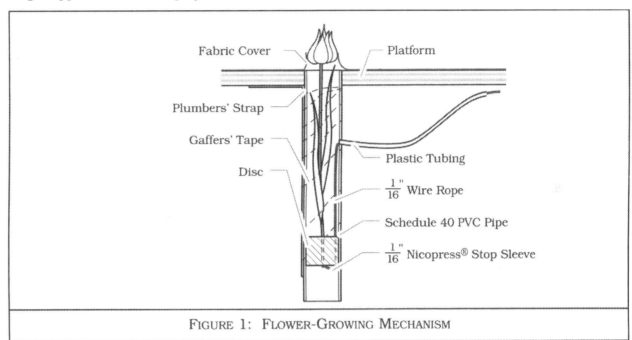

FIGURE 1: FLOWER-GROWING MECHANISM

MATERIALS

> 2' pieces of 1" to $1\frac{1}{2}$" schedule 40 PVC Pipe
> $\frac{1}{2}$"-thick Disc of Wood for the piston
> $\frac{1}{16}$" Wire Rope
> 3' piece of Plumbers' Strap
> $\frac{3}{16}$" Plastic Tubing
> Gaffers' Tape
> $\frac{1}{16}$" Nicopress® Stop Sleeve
> Artificial Flower of Your Choice

The mechanism described below requires 9" to 15" of space below the deck surface. There should also be a great deal of space available below the deck for easy access and installation. If you are using multi-flowered stems, or would like more than one flower to grow from a stem, I recommend using a larger-diameter PVC pipe.

Cut the PVC to a length that is about 3" longer than the flowers you plan to use. If you are working with a raked deck, be sure to cut the PVC at the angle of your deck. Determine the thickness of the deck at the place where the flower will be placed, add three inches for working clearance, measure and mark this distance from the top of the pipe, and just below this line, drill a hole large enough to allow the plastic tubing to fit through.

Next, you need to create a holder for the flower in the pipe. Cut a disc out of the $\frac{1}{2}$" piece of wood that will fit the inside of the PVC pipe with enough room to slide up and down the pipe easily. The ease with which the disc slides within the pipe is crucial to the operation of the effect. Sand the edges of the disc well and cover them with a non-greasy lubricant when the mechanism is ready to be put together. Drill a hole large enough for the wire rope to pass through into the center of the disc. Now drill a hole for the stem of the flower.

Cut a hole to accommodate the outside diameter of the PVC pipe into the deck where the flower is to grow. Next, insert the wire rope through the plastic tubing. A little graphite squirted into the plastic tubing will ease the insertion. Install the plastic tubing below the deck between the point from which the effect will be run and the site of flower growth. At this time attach a stop to the off-stage end of the wire rope to prevent it from slipping through the plastic tubing while you work.

On the onstage end, slip the plastic tubing and wire through the hole in the PVC and let it extend out of the bottom of the tube. Insert the PVC through the hole in the deck until its top is flush with the top of the deck. Screw the plumbers' strap to the bottom of the deck and bend it 90° so that it is adjacent to the PVC. Using gaffers' tape, tightly secure the PVC to the plumbers' strap. The PVC should now be rigidly secured to the deck.

The remainder of the installation is best accomplished by two people. Insert the onstage end of the wire rope through the appropriate hole in the disc. Attach the Nicopress® stop sleeve to the end of the wire rope to prevent it from slipping back through the hole. Spray the sides of the disc with Pledge® or a similar furniture wax so it will slide through the PVC. The plastic tube can now be pulled out of the PVC until it is just inside the hole in the side of the pipe. This is a good time to check the action of the mechanism. By pulling on the offstage end of the wire rope, the disc should easily slide up the PVC pipe. Since it is not easy to push the wire back through the tube, it is necessary to go onto the deck with a rigid piece of wire or wood and push the disc back down the tube from above.

Once the system is working satisfactorily, the flower can be added to the mechanism. The stem should be cut about 2" shorter than the PVC. Push the disc out of the bottom of the tube again. Have one person insert the flower into the tube from above until the person below can grasp the bottom of the stem. The stem should then be inserted into the hole in the disc, bent over, and stapled or otherwise secured to the bottom of the disc. The first person should then pull again on the wire while the disc is guided into the bottom of the tube. When the disc is approximately 1" above the bottom of the tube, a second mark should be made on the offstage end of the wire. This will prevent the disc from being pushed out of the tube during the reset of the flower. With a gentle pull on the wire rope, the flower should emerge from the surface of the deck.

To cover the hole in the deck surface, cut two small pieces of fabric and attach them to the surface of the deck. Each piece should cover half of the hole. The attachment of the fabric should leave it loose enough to allow the flower to push through with little difficulty. Paint the fabric to match the surface of the deck.

If your show requires that the flower or flowers be pickable, glue a piece of drinking straw perpendicular to the top of the disc instead of drilling a hole through the disc. The flower stem can then be inserted into the straw, and pulled out at any time. Resetting of the effect in this case requires a little more care to assure that the flower stays seated in the straw as it is pushed into the tube.

While he was Production Manager at the Guthrie Theatre, Terry Sateren developed a simple and easily repeated process for making panes of break-away glass. The process uses a pane of plate glass to make a negative mold for a break-away casting of hydrocarbon thermoplastic resins. It requires very simple tools and produces consistent results.

TOOLS AND MATERIALS

> Refrigerator
> Stove or oven
> Electric saucepot
> Heat-conducting support plate (we used a $\frac{1}{4}$" aluminum plate larger than the final mold)
> Cooling plate (we used a $\frac{1}{2}$"- or $\frac{3}{4}$"-thick section of glass plate, larger than the final mold)
> Level cooling stilts (wood or metal)
> Flexible mold
> Pico® plastic(s)

THE MOLD

For a glass pane, the best results (both for sizing and surface quality) are produced by making the mold from a section of plate glass. The mold must be flexible in order to peel away from the casting, and must be able to withstand heat and cold with a minimum of temperature-related expansion and contraction. We chose Dow Corning Silastic® "E" RTV rubber with a Shore A durometer hardness of 35, tensile strength of 700psi, tear strength die of 90ppi and a 400% elongation specification.

For a production of *The Misanthrope* we built the mold pictured in Figure 1, which produced a break-away glass pane with a thickness of $\frac{1}{4}$".

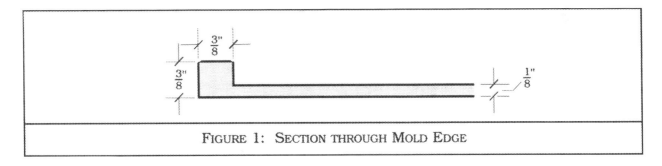

FIGURE 1: SECTION THROUGH MOLD EDGE

THE RESINS

The actual break-away glass is composed of a two-part mix of hydrocarbon thermoplastic resins. For our break-away glass we used a mixture of 75% Piccolastic D100 Resin® and 25% Piccotex 100 Resin®. The former resin is amber colored; the latter, white. This proportion produced a transparent, amber-tinted pane that broke remarkably well, producing good-sized shards and a realistic sound. The resins are available from

> Hercules Incorporated
> Hercules Plaza
> Wilmington, DE 19894

These plastics are available in a solid form (400-lb. drums), and flake form (50-lb. bags). We used the flake form for all castings. The flakes must be kept away from heat, and used within a reasonable amount of time, or they tend to fuse, block, and lump up. Old flake material may produce a darker product (because of oxidation) or may not mix at all. Moreover, the flake form does get crushed down, producing a fine "sand" or particulate that is very flammable when suspended in air. Keep the bags away from sparks, open flames, and high temperatures.

THE PROCESS

Once the mold had been produced, we followed a specific process to cast the break-away panes. This process entailed melting the mixture of resins, pre-heating a support plate, pre-chilling a cooling plate, and pouring the resin mix very carefully. A heated support plate heats the mold and resin mix evenly at pour time, and a chilled cooling plate speeds the setup time and helps insure that the casting will cool evenly. Placing the cooling casting on level stilts ensures a consistent thickness throughout the casting and allows air to circulate around the casting for cooling. Following is a detailed step-by-step approach to producing a pane of break-away glass. You may need to adjust it to suit your particular application.

1.   Melt 75% Piccolastic® and 25% Piccotex® resins in an electric saucepot. Mix resins evenly as you pour them into the pot. Let the mixture melt together for one hour.

2.   Heat the RTV mold on the support plate in an oven for approximately 15 minutes.

3.   Chill the glass cooling plate in a refrigerator for one hour.

4.   Evenly and carefully pour the melted resin mix into the pre-heated RTV mold. Take care here not to pour too fast or drip to avoid causing air bubbles to become trapped in the casting. Let the mold and heating plate cool inside the oven on a level rack for 15 minutes.

5.   Carefully slide the mold off of the heating plate and onto the cooling plate.

6.   Place the cooling plate and mold on level stilts and let cool at room temperature for approximately one and one-half hours.

7.   Carefully peel the mold off the break-away glass pane and store the casting in a safe, padded place.

NOTES

We made at least two panes for every one needed for the production since accidents happen and some castings suffer from poor mixing or bubbles. The proportion of the resins to one another will vary the color, brittleness, rigidity, and clarity of the break-away product. Like other thermoplastics, the resins are theoretically re-usable: it should be possible to re-melt the shards and form them into new panes. Since we chose to use new resins for each glass pane, the number of times this could be done was not determined.

ᘿᛁᘿᛁᘿᛁ

For a Yale School of Drama production of Shakespeare's *Measure for Measure* the director asked if we could build a set of torches for the actors to carry. Because of the strict fire laws in Connecticut we discussed the possibilities with the Fire Marshal and came to the conclusion that it would be safer to build torches that used a butane canister than it would be to use Sterno®. Early on we decided that we should be able to control the flame height and that the torches should go out if put down or dropped. The technical design of the torch was in part determined by the intended look, but can be adapted for other applications.

The part numbers in Figure 1, which illustrates the mechanism, are Clippard catalog numbers. Not all parts are shown, however: see the parts list at the end of this article for a complete inventory.

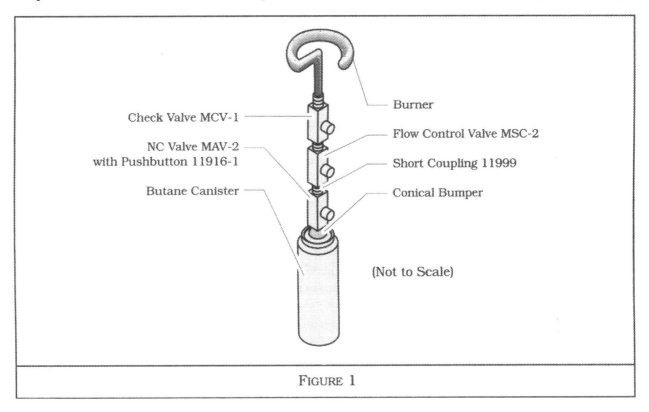

Check Valve MCV-1

NC Valve MAV-2
with Pushbutton 11916-1

Butane Canister

Burner

Flow Control Valve MSC-2

Short Coupling 11999

Conical Bumper

(Not to Scale)

FIGURE 1

One of the biggest problems at the beginning was how to achieve a good seal between the hose feeding the burner and the neck of the butane canister. The Clippard miniature fittings and valves we wanted to use terminate in threaded ends. We discovered that the nozzles supplied with most butane canisters are not durable enough to be threaded. After playing with a number of fittings we discovered that the small conical plastic bumpers used as feet on electronic circuit cabinetry fit snugly over the nozzle. By threading a Clippard fitting into the top of the bumper we achieved the necessary firm connection. It is very important when assembling the various connections to attach the hose firmly and seal all the threaded connections with some type of thread-lock compound. There must be no leaks.

The next problems involved provisions for a dead-man switch for the gas and for a method of adjusting the flow. We solved both problems with Clippard Minimatic valves. For the dead-man switch we used a spring-loaded normally closed valve operated with a $\frac{1}{2}$"-diameter pushbutton that worked well with our design and could be easily hidden. We controlled the flow of gas with a miniature needle valve, and for safety we added a miniature check valve to keep any burning gas from backing up should anything go wrong. For the burner we used a piece of $\frac{1}{4}$" copper tubing bent into a circle with nine $\frac{1}{16}$" holes drilled in it.

Though regular $\frac{1}{4}$" vinyl hose used with Clippard compression fittings would probably be safe, we used rubber hose rated for gas. Although we never timed it, the medium-sized canisters of butane lasted at least forty minutes.

PARTS LIST

1	MAV-2	Normally Closed 2-Way Valve
1	MCV-1	Check Valve
1	MSC-2	Adjustable Flow-Control Valve
1	11916-1	Pushbutton
2	11752-2	10-32 to $\frac{1}{16}$" Hose Barbs
1	15010	Extension Fitting
1	11999	Short Coupling
1	15006-2	Pipe-to-Female Adapter
2	5000-2	Hose Clamps
		Vinyl Hose
		Bumper for electronics cabinetry

Hand-held live flame torches should meet certain safety guidelines while also being both inexpensive and simple to construct and operate. Propane or butane gas can be incorporated into good torch designs (see "A Butane Torch for Use Onstage" by Rod Hickey), but a local ordinance preventing the use of bottled gases onstage forced a different approach. The design presented here uses lamp oil as a fuel, is self-extinguishing if dropped, has a proven burn time of at least ten minutes, and is based on technology only as complex as a candle snuffer.

Basically, one cylindrical chamber holding an oil-soaked wick slides within another cylinder. So long as the torch bearer holds onto a knob that keeps the inner chamber in its "up" position, the wick is exposed and can be lit. If the torch is dropped, however, a spring retracts the inner cylinder and wick past a damper that flips down and seals out the air needed to maintain burning.

Construction consists mainly of machining two pieces of telescoping pipe. The inner pipe, an 8" piece of $1\frac{1}{2}$" chromed brass sink drain pipe into whose lower end a $\frac{1}{4}$"-thick brass disc has been soldered, forms the housing for an oil-soaked wick. Since even a pinhole would have allowed excess lamp oil to drip out the bottom of the torch, I used a metal lathe to turn the disc to a uniform 1.450" diameter. A (roughly) $\frac{1}{2}$"-deep hole drilled through the pipe and into the side of the disc and tapped for a 10-32 machine screw provided a mounting point for the wick chamber's knob — a 1" machine screw with several nuts jammed up on the head.

The second piece of pipe forms the body of the torch. In our version it was an approximately 14"-long piece of $1\frac{1}{2}$" schedule 40 aluminum pipe. Schedule 40 black pipe could be used, of course, but its heaviness would be a considerable disadvantage. This second pipe is machined in three places. A $\frac{1}{4}$"-wide slot is cut along the length of the pipe between points $7\frac{3}{4}$" and $9\frac{3}{4}$" from one end to accommodate the knob movement. On the opposite side, and $1\frac{3}{4}$" from the same end, a $\frac{1}{2}$" hole and two smaller holes are drilled as shown in Figure 1 to provide a mounting for the damper pin and spring. Lastly, a small hole is drilled at a 45° angle into the end opposite the $\frac{1}{2}$" hole as a mounting for the retracting spring.

A spring is used to retract the inner pipe when the knob is released. One end of this spring hooks onto a small wire loop soldered to the bottom of the wick chamber. The other end attaches to the outer pipe at the small hole drilled at one end. The spring should be strong enough to pull the inner pipe even if the torch is held upside down, yet not so strong that its pressure is uncomfortable to the torch bearer.

The part that actually snuffs out the flame, the damper, is a piece of sheet metal cut into the shape shown in Figure 1 and bent to the radius of the inside of the outer pipe. (Though its shape is, theoretically, formed by the intersection of two equal-radius cylinders at a 45° angle, this damper is most easily made through trial-and-error fitting and adjusting.) When the torch is burning, this piece must fit between the inner and outer telescoping pipes. When the inner pipe retracts, the damper, pushed by a small spring, flips down to seal out air.

Wicks of both cotton and Fiberglas® fabric have been used. The Fiberglas® is superior if the burn time exceeds a few minutes because the cotton does not seem to be able to wick up fresh oil fast enough to keep itself from burning up.

DESIGNED WITH

Corky Boyd.

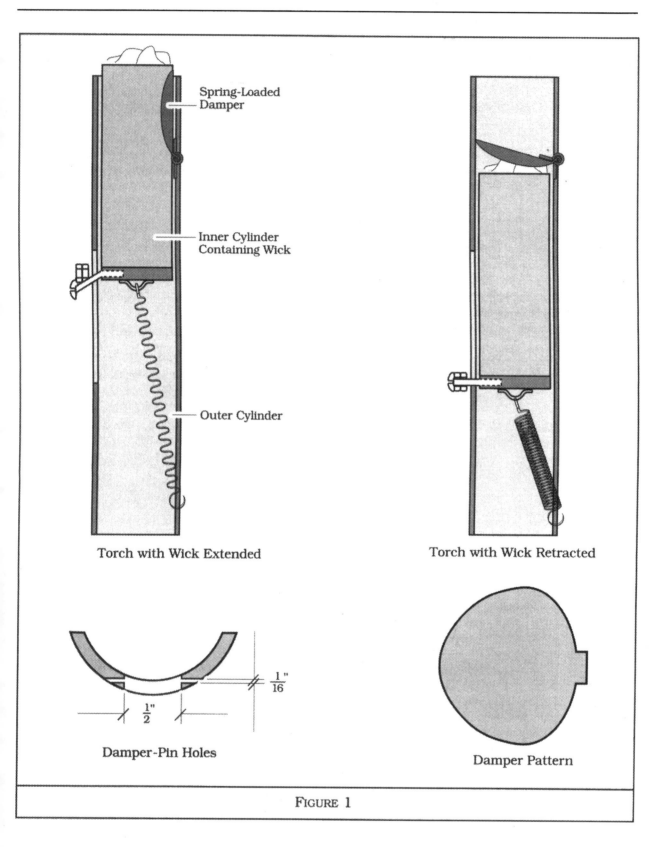

Spring-Loaded Damper

Inner Cylinder Containing Wick

Outer Cylinder

Torch with Wick Extended

Torch with Wick Retracted

$\frac{1}{16}$"

$\frac{1}{2}$"

Damper-Pin Holes

Damper Pattern

FIGURE 1

Battery-operated practicals whose performance can be remotely controlled are certainly not new. Like others, the wireless practical described in this article can be brought onstage fully lit, left unattended, and will dim out as the result of a fade to black. But, rather than being controlled by a dimmer, this particular design uses a photoresistor to make it fade up and down in response to the changes in the ambient stage lighting: as the stage fades up, so does the practical; and as the stage goes to black at a scene's end, this practical goes out as well.

The schematic in Figure 1 illustrates the simplest prototype of this device, which was designed as a class project. The basic components are readily available at electronics hobby stores like Radio Shack, and construction of the unit is fairly straightforward. The only caution worth noting is that the photoresistor must be shielded from the spill of the 12V lamp, or the practical's own brightness will keep the photoresistor continuously activated and the circuit will stay on.

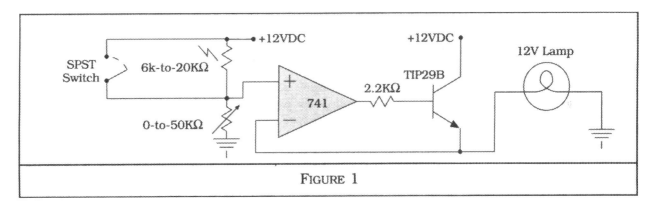

FIGURE 1

PARTS LIST

> 1 6 to 20KΩ Photoresistor (such as EG&G Vactec's VT-333, dark resistance 10MΩ; 2 fc resistance = 28KΩ)
> 1 741 Operational Amplifier (op amp)
> 1 0 to 50KΩ Variable Resistor
> 1 2.2KΩ Resistor
> 1 TIP29B Transistor (NPN)
> 1 12V Lamp
> 1 SPST Switch

In addition to the essential components, the prototype features a variable resistor to adjust the overall range of brightness, and an override switch that disables the sensor so that the practical can be carried, lit, onto a dark stage. A more sophisticated version might include additional photoresistors wired in parallel with that shown. The use of multiple photoresistors would eliminate the undesirable dim-out that would occur in the prototype if its single sensor were accidentally obscured.

SOURCE

Alan Hendrickson.

Faking Waterproof Beer Can Labels

Chris Higgins

For a production of *The Beach* at the Yale Repertory Theatre, I was faced with the familiar problem of providing faked beer cans for use onstage. The cans had to fizz when opened, and had to be waterproof, as they were to be stored in a cooler full of ice.

I considered two solutions to this common problem before making the choice detailed in this article. The first was to drain real cans from the bottom, refill them with some non-alcoholic liquid, and then seal them with a waterproof adhesive tape such as duct tape. This approach was out of the question because, in addition to being fairly costly, it would have required far too much pre-show setup time. The second alternative, that of covering soda cans with the generic, self-adhesive labels used in film and television or with color Xerox prints of real labels, seemed promising. But our designer and director wanted real labels, and finding a way to waterproof paper labels seemed potentially troublesome.

The solution I chose was to cut a real beer can into a "label" to be applied to a soda can. Simply removing the top and bottom of an actual beer can and making a vertical cut along its back turns it into a label that can be repeatedly wrapped around any same-size soda cans. I used Budweiser® beer cans, because the label design is printed on a white background that is easily simulated by white tape.

Though each can's top and bottom could be removed with a hacksaw, I used a metal-cutting bandsaw and took steps to avoid crushing the can in the process. I cut off the top first, since the tops resist crushing better than the molded bottoms. Then, before cutting the bottom, I reinforced the can by inserting a short length of pipe through its open end. Next, having slit the can up the back with a pair of tin snips, I trimmed all of the edges on a sheet-metal shear, making sure that a $\frac{1}{8}$" border remained around the label's printing to allow for attaching the label to the can of soda. I then used a metal file to smooth each label's edges. The labels took about ten minutes each to make.

The finished labels were first taped to the cans along the vertical seams at the back, and then wrapped around the cans and taped in place along both the top and bottom edges. White plastic electrician's tape worked very well because it blended in with the Budweiser® label background and stretched to conform to the shape of the soda can, completing the illusion. The labels I made withstood repeated attachment to soda cans during a four-week run and showed no appreciable signs of wearing out.

NOTE

Self-adhesive labels are available in a variety of copyright-free designs from

The Earl Hays Press
10707 Sherman Way
Sun Valley, CA 91352
(818) 765-0700

Casting Bondo® (autobody filler) in molds made from plasticine proves a quick and inexpensive method of reproducing small, low-relief objects. It is a faster alternative to the more common approach of casting liquid latex in plaster molds. The Bondo® technique was developed in the process of reproducing the finials of an existing brass headboard. It was impractical to make latex casts because of the time commitment involved.

Latex casting requires a great deal of linear time because the initial plaster mold must dry completely before it is used, and each latex cast poured can take a full day or more to dry. Thus, making a plaster mold and a latex cast of a deep object can take three to four days. In contrast, plasticine and Bondo® casting yields a rigid product ready for painting in under two hours.

SUITABLE ORIGINALS

This technique works well for recreating cast metal objects or other objects that were originally made by a casting process. The object to be reproduced should have no undercuts since the mold will distort when it and the object are separated. Objects that are more than half round cannot be cast as one piece. To recreate full-round objects, cast each half separately and join the two halves.

MAKING A PLASTICINE MOLD

Plasticine responds to temperature, softening when warmed and stiffening when chilled. Since it is portable and cohesive, unlike wet plaster, plasticine molds of gate and door hardware and the like can be made away from the shop. Also, plasticine is non-reactive and will not damage non-porous originals.

Begin with a lump of plasticine large enough to cover the object $\frac{1}{2}$" to 1" deep, *i.e.,* deep enough to keep the mold from distorting. Work the plasticine with your hands until it is soft. Form the lump into the general size and shape of the object. Press the plasticine firmly onto the object; do not pull it around the object with the fingers, as this can smear the image of the mold. If possible, use a refrigerator, an air conditioner, ice packs, or even the temperature of a winter's day to chill the mold and the object before attempting to separate them, for stiff plasticine will better resist distortion. If chilling is impractical, let the mold and object sit for a while so that as much heat as possible dissipates into the air. Next, carefully peel the two apart, trying not to bend or twist the mold. Attempt to true the mold if separation has produced any mild distortion. If severe distortion has occurred, rework the plasticine and try again. Each mold will probably make only one cast, and making more molds than you will need is advisable and will take very little time. When you've finished making molds, use soap and water to remove the oily residue.

MAKING A BONDO® CAST

Once the mold is ready, mix the two parts of the Bondo® following standard instructions. Pour the Bondo® into the mold immediately. No additional release agent is needed since the oil in the plasticine serves this purpose. To avoid air bubbles, use a stick to work the Bondo® into the deepest areas first. In three to ten minutes, the Bondo® should be hardened. Carefully pull the cast from the mold. Though (with great care) it may be possible to use a mold more than once, the heat of the chemical reaction that hardens the Bondo® softens the plasticine, making it difficult to remove the cast from the mold without distortion. While the cast is still green, *i.e.,* hardened but not completely cured, trim away excess material with a matte knife, rasp, or sandpaper.

NOTES

The detail achieved in this process may not be as fine as casting latex in plaster, but for most the-atrical purposes it is sufficient. Besides, it is very fast and inexpensive when compared to latex casting. Using this technique, I copied twenty-two finials for the headboard in four hours, and cre-ated a pair of matching period door plates from one original in less than forty-five minutes. It is possible to make many plasticine molds and Bondo® casts of a single original in the time it takes one plaster mold to dry. As for the expense, a gallon of Bondo® costs less than one-quarter that of an equal volume of latex. Its last advantage over latex casting is speed. Because Bondo® casting involves a chemical reaction, casting time is independent of environmental factors, while the evap-oration of latex is greatly hampered by humidity.

In fact, there is only one disadvantage to this process: using Bondo® poses a health risk. Read the health warnings printed on the container and take the necessary precautions.

❧❧❧

For a production of *A Funny Thing Happened on the Way to the Forum*, I needed to coat a practical water fountain sculpture that was manufactured in our shop with a substance that would waterproof the porous plaster and foam sculpture, add some rigidity to the draped fabric parts, and be both thick enough to fill in gaps but thin enough not to cover some of the finer detail. My first inclination was to use auto body compound available at any discount department store or auto supply shop. The putty by itself was too viscous to be applied to the sculpture and keep the detail clear. Fiberglas® resin by itself was too thin, and ran and dripped off the form. However, by mixing the body compound with Fiberglas® resin made by the same manufacturer, I was able to liquefy the putty enough so that it could be painted on the sculpture with a stiff-bristled, disposable brush. The foam parts also had to be covered with plaster bandage in order to keep from cavitating under the polyester mixture.

Through a little research, I discovered that even though brand names may differ, most commercial auto body putty and Fiberglas® resins have the same polyester base and can compatibly mix. The catalyst for either material can be used to activate the hardening process. Setup time can be speeded up by adding more catalyst or by stirring the mixture for a longer period of time.

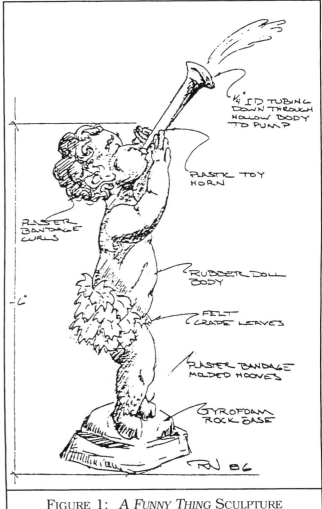

FIGURE 1: *A Funny Thing* Sculpture

EDITORS' NOTE

Working with auto body putty and Fiberglas® resin presents a known health hazard. Wear protective clothing and gloves and an appropriately filtered respirator while using these substances.

❧❧❧

A Very Useful Snow Machine

Richard Gold

In the real world, the soothing effect of snow falling gently behind a window is one of nature's gifts to those who must also deal with the consequences of inches of accumulation the next day. But for the Yale Repertory Theatre's production of *Ohio State Murders*, the effect of snow was not nature's gift but a show requirement. The parameters additionally included that the snow be concentrated in a 3' width upstage of a window hanging in free space, and of course that the mechanism be as quiet as possible.

For "snow," the production's propmaster selected and purchased Consolidated Display's Scatter Flakes® on the basis of looks and fire-resistance. Upon inspecting it, we concluded that the mechanism constructed would have to be as jam-proof as possible, since the material seemed easily capable of jamming in an opening and preventing further snowfall. This factor alone eliminated many of the devices we had been considering: snow bags, chutes, conveyors, and auger mechanisms. Eventually, we determined that a rotating drum could be the solution to the problem.

We bought a 30-gallon fiber drum from a nearby barrel distributor for $15.00 and modified it as shown in Figure 1. A compact, split-capacitor motor with a gear reducer that would provide an output of about 4 rpm and 80 foot-pounds of torque was available and seemed perfectly suited for the job — not too noisy or bulky. We chose a 3:1 gear ratio to meet the necessary speed and torque requirements. After we assembled the parts on a simple bracket, we tested the apparatus. Through trial and error, we determined an appropriate hole size (1") and pattern to perforate the drum, which allowed the snow to fall at an appropriate pace and density. This approach had the

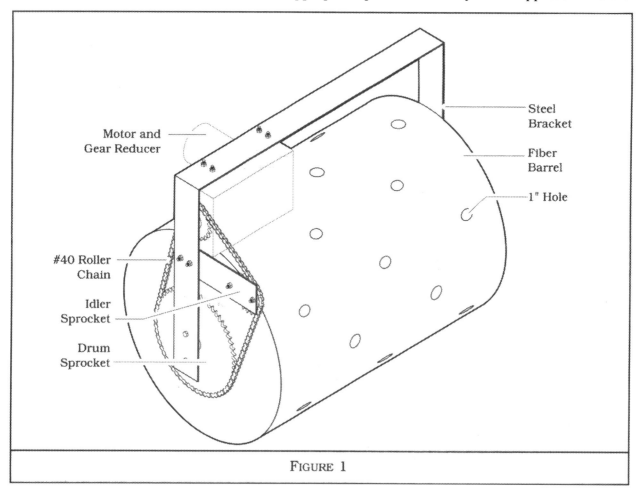

Motor and Gear Reducer

Steel Bracket

Fiber Barrel

1" Hole

#40 Roller Chain

Idler Sprocket

Drum Sprocket

FIGURE 1

additional advantage that if the designer or director wanted the snow to fall at a different speed, additional holes could be drilled or existing holes could be taped closed.

We felt certain that there would be no problem providing a snowfall for the one-hour duration of the show, but just how long a barrelful of snow would last in 8- to 12-hour techs was a point of some concern. Our estimates proved to be conservative, as a half-full barrel lasted approximately 6 hours, and the operator was instructed to turn the machine off whenever extended periods were to be spent looking at other aspects of the show.

The only problem we encountered was the mysterious loosening of the roller chain after a week of smooth operation. With the addition of an idler sprocket, trouble-free operation was restored and the machine worked fine through the run of the show.

The snow machine met all the requirements of the production, including quiet operation, predictable patterns of snowfall, simple construction and installation, and budgetary constraints.

Thanks to Patricia Bennett for the basic idea and M. Craig McKenzie for his help. Scatter Flakes® are available from Consolidated Display Company, Inc., Naperville, IL.

❄❄❄❄

Rigging Hardware

A Hinge Device to Facilitate Hanging Vertical Booms   *William R. Wyatt, Jr.*

The hinge device shown below can save time and facilitate accurate counterweighting of vertical booms.   See Figures 1A, 1B, and 2. The device makes it possible to attach a boom to a batten, load the arbor with the batten in, and then fly the batten out.

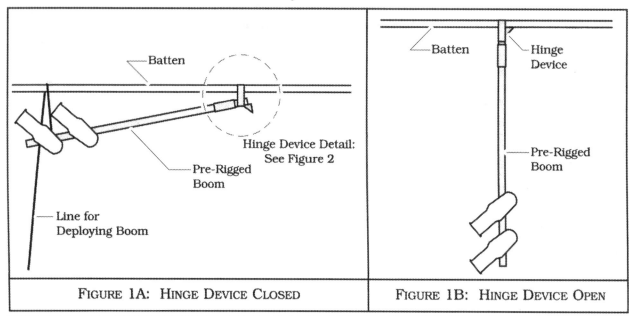

FIGURE 1A: HINGE DEVICE CLOSED   FIGURE 1B: HINGE DEVICE OPEN

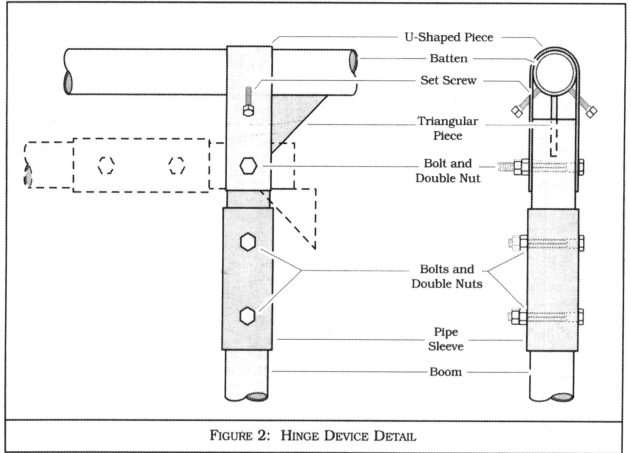

FIGURE 2: HINGE DEVICE DETAIL

MATERIALS

The main components are as follows:

1. A U-shaped piece of strap steel bent to fit over the batten.

2. A piece of schedule 40 black pipe 1' long and the same diameter as the batten with a triangular piece of steel welded to one end. The triangular piece serves as a stop keeping the boom vertical.

3. A larger-diameter black pipe sleeve to join the 1'-long piece to the boom.

4. Double-nutted bolts to hold the pieces together.

ASSEMBLY

These are the steps to attach the boom to the batten using this device:

1. Fly the batten in.

2. Put the U-shaped piece over the batten and tighten the set screws.

3. Sleeve the pipe sections together.

4. Place the boom parallel to the batten and connect the pipe to the U-shaped piece with double-nutted bolts.

5. Attach the unhinged end of the boom to the batten so it clears the floor. The hinge device is now in the closed position. See Figure 1A.

6. Load the arbor.

7. Release the unhinged end of the boom and fly the batten out while lowering the boom into a perpendicular position. The hinge device is now in the open position. See Figure 1B.

NOTE

When the hinge device is in the open position, gravity holds it in place. Therefore, most of the weight on the boom must be on the side opposite the triangular piece.

DESIGNED WITH

Richard Thurman.

ゑ▲ゑ▲ゑ▲

Sandwich Batten Clamps

Jon Lagerquist

This clamp is a low-cost means of attaching sandwich battens to a drop or scrim. It was designed to stretch a scrim folded on its centerline without puncturing the scrim or using all of the shop's C-clamps. The clamp consists of the four parts shown in Figures 1 and 2:

A   Clamp Body. $\frac{3}{4}$" plywood. The shape of the notch and the width of the arm are the critical elements.

B   Sandwich Battens. Two 1x3s the length of the drop.

C   Side Wedges. Two 4" pieces of 1x3 cut with a bevel to match the notch in the clamp body.

D   Bottom Wedges. Two 8" pieces of 1x3 tapered to a point.

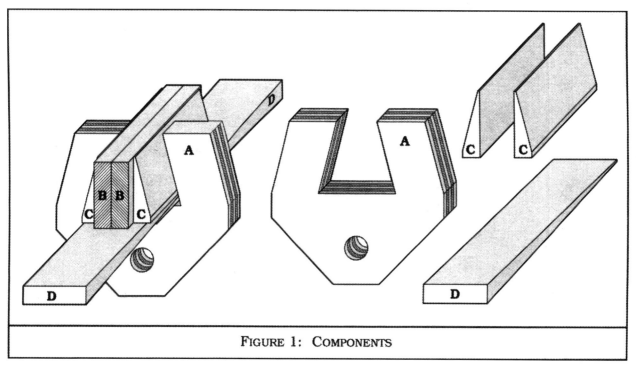

FIGURE 1: COMPONENTS

INSTALLATION

1. With a contrasting color thread, baste a line on the drop where the top of the batten should be.

2. Enclose a fold of the drop between the two halves of the batten. See Figure 2. Do not allow any of the drop to extend beyond the bottom of the batten because the drop could be damaged during the tightening process.

3. Temporarily secure the battens with C-clamps.

4. Slide the clamp body over the battens.

5. Insert the side wedges.

6. Insert the bottom wedges, one from each side.

7. Drive the bottom wedges together to tighten the clamp.

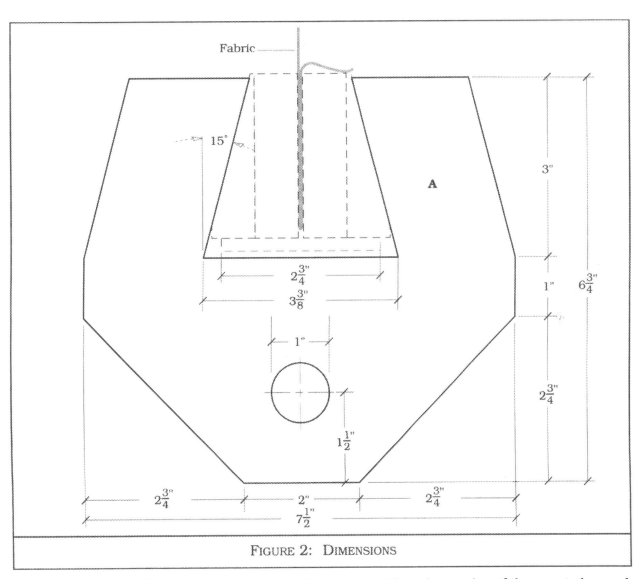

Place the clamps on 4' centers. In most cases the clamps will not loosen, but if they are to be used overhead, a few staples should be used to secure the bottom wedges to each other.

Spider-Cable Clamps

Tom Neville

The "spider," a set of multiplexed outlets attached to a long, grid-mounted cable, allows an electrician to use any batten as an electric. Once such a new electric has been flown to the desired trim, however, the electrician is left with the problem of securing the excess cable at the grid. This article describes a quick and safe solution to that problem — the spider-cable clamp. The spider-cable clamp consists of two pieces of 2x4 joined at one end by a bent strap hinge, and at the other by a Roto Lock® mounted on pieces of steel flat stock. See Figures 1 and 2.

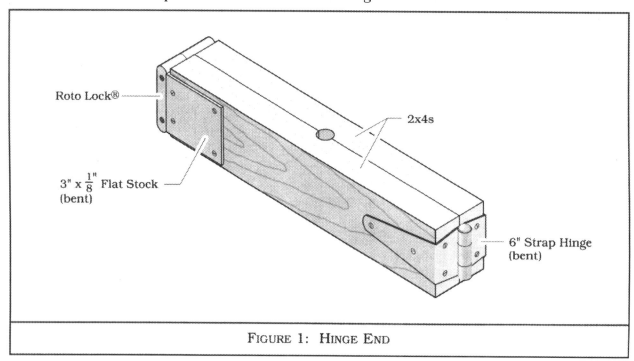

FIGURE 1: HINGE END

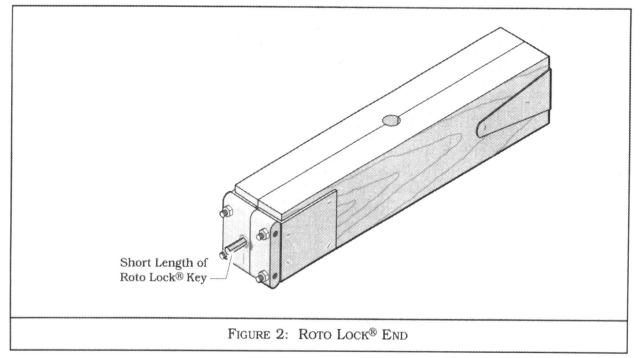

FIGURE 2: ROTO LOCK® END

CONSTRUCTION

1. Cut two pieces of 2x4 into lengths that will safely span the grid wells. The author advocates the use of hardwoods such as maple or oak.

2. Clamp the 2x4s together face-to-face and drill a hole along the plane of contact between them and equidistant from both ends. Make the diameter of this hole $\frac{3}{16}$" smaller than that of the cable to be clamped.

3. Bend a 6" strap hinge to wrap around one end of the assembly. Attach it with 1" #10 screws as shown in Figure 1.

4. Bend two pieces of 3" x $\frac{1}{8}$" steel flat stock to wrap around the other end of the assembly as shown in Figure 2.

5. Hold the locked Roto Lock® in place and mark the location of the Roto Lock® bolt holes on the steel. Accuracy in marking is especially important during this step.

6. Remove the steel from its clamps, and drill each Roto Lock® bolt hole $\frac{1}{16}$" closer to the right-angle bend to assure adequate clamping compression. Drill four screw holes in the long leg of each piece of steel. Countersink the bolt and screw holes.

7. Braze a short length of Roto Lock® key into the keyhole to eliminate the need for taking any tool other than a safetied crescent wrench aloft.

⛄⛄⛄

This system can be used in place of a commercially available traveler track to move light scenery pieces across the stage. Lacking the weight capacity of roller systems, it is nevertheless much quieter and less expensive, and it can be made from readily available materials.

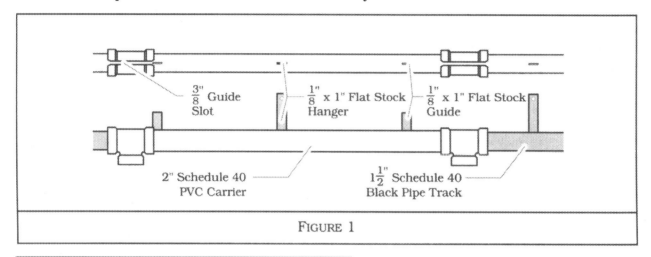

$\frac{3}{8}$" Guide Slot

$\frac{1}{8}$" x 1" Flat Stock Hanger

$\frac{1}{8}$" x 1" Flat Stock Guide

2" Schedule 40 PVC Carrier

$1\frac{1}{2}$" Schedule 40 Black Pipe Track

FIGURE 1

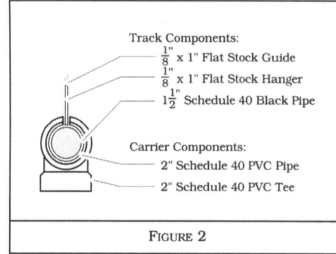

Track Components:
$\frac{1}{8}$" x 1" Flat Stock Guide
$\frac{1}{8}$" x 1" Flat Stock Hanger
$1\frac{1}{2}$" Schedule 40 Black Pipe

Carrier Components:
2" Schedule 40 PVC Pipe
2" Schedule 40 PVC Tee

FIGURE 2

The basic elements of the system are the carrier and track illustrated in Figures 1 and 2. The carrier is made from 2" schedule 40 PVC pipe with a 2" PVC Tee fitting at each end. To prevent the scenery from racking laterally if bumped or moved quickly, the carrier should always be as long as the scenery it carries is wide. After the PVC pipe has been cut to the desired length, the Tee fittings are cemented onto each end with PVC cement. When the cement is dry a $\frac{3}{8}$" slot is cut along the entire length of the carrier using a dado blade on a table saw. The slot is further widened at each end of the carrier to help align it with the hangers and guides of the supporting track.

The supporting track is made from $1\frac{1}{2}$"-diameter schedule 40 black pipe. If runs over 21' long are needed, the pipe must be joined with an inside sleeve so that the entire run is smooth. Flat stock measuring 1" x $\frac{1}{8}$" is used as hangers for the track and as guides for the carrier. The hangers we used were 4" long with a $\frac{3}{8}$" hole at one end, and the guides were 2" long. We placed a hanger every 8' along the pipe to provide support. To prevent the carrier from rotating on the track, the distance between any two hangers or guides should not exceed the length of the PVC carrier.

After the carrier is assembled, it is slid onto the track and the scenery is attached to its Tee fittings. The scenery can be moved either by an onstage actor or by an offstage crew using a conventional sash cord and pulley system. Ordinary bar soap works well as lubrication between the carrier and the track, and has the advantage of not dripping or staining.

Although theatrical rigging often requires that multiple pipes be spliced together in order to form a continuous batten, the most common techniques for achieving such splices are either dangerously ineffective or unnecessarily involved. Threaded couplings are too weak to be used for structural splices, and internal cylindrical sleeves are difficult to fit and secure properly. The Channon Corporation (now Chanco, Ltd.) recognized these problems, and, in the sixties, developed the internal expansion splice (IES) as a fast, neat, and effective means of splicing pipe.

The IES consists of two adjacent steel bars that are held in alignment by roll pins and fitted with set screws that determine the distance between the bars. After the pipes are butted together with the splice straddling the joint, the splice is expanded, thereby locking itself and both pipes in place. The IES is superior to other splices since it is extremely resistant to bending while not requiring complicated machining processes or welding. Other advantages of the IES are its complete internalization (no protrusions from the batten's surface), and its ease of removal.

The splice pictured in Figures 1 and 2 is sized for $1\frac{1}{2}$" schedule 40 black pipe, the most common material used for stage battens. By using smaller or larger cold-rolled steel bars, however, the splice is easily adapted to fit any pipe commonly used in theatre. If a splice of another size is to be fabricated, the length of the steel bars should remain 1', while their width and thickness should be varied in accordance with the pipe's inside diameter. Be aware that pipes of the same nominal size yet of differing schedules will usually require different-sized splices.

MATERIALS LIST FOR ONE $1\frac{1}{2}$" SCHEDULE 40 IES

2 pieces	$\frac{1}{2}$" x $1\frac{1}{4}$" Cold-Rolled Steel
2	$\frac{5}{16}$" x 1" Roll Pins
2	$\frac{3}{8}$" x 1" Set Screws

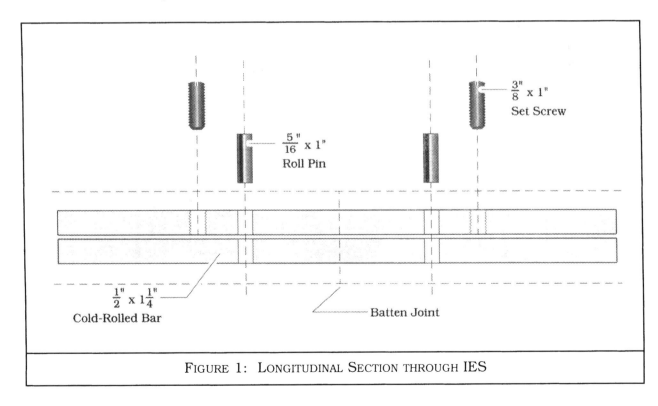

$\frac{3}{8}$" x 1" Set Screw

$\frac{5}{16}$" x 1" Roll Pin

$\frac{1}{2}$" x $1\frac{1}{4}$" Cold-Rolled Bar

Batten Joint

FIGURE 1: LONGITUDINAL SECTION THROUGH IES

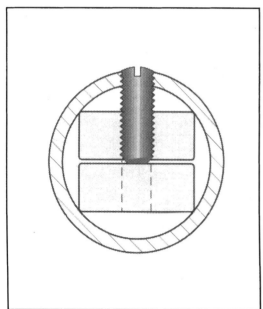

FIGURE 2: SECTION THROUGH IES AND BATTEN AT SET SCREW

CONSTRUCTION AND INSTALLATION

No matter what size splice is to be fabricated the procedure is identical. First, cut two 1' pieces of steel bar and drill a $\frac{5}{16}$" hole 4" from each end of both pieces. Next, set one piece aside and, in the other, drill and tap for a $\frac{3}{8}$" set screw 3" in from both of its ends. Finally, stack the two steel bars so as to align the pairs of $\frac{5}{16}$" holes, and drive a roll pin into each pair of holes in order to connect the two steel bars.

Ease of installation is one of the IES's greatest advantages. To install, drill a $\frac{1}{2}$"-diameter hole 3" in from the end of each pipe to be spliced. Insert the IES halfway into one pipe and rotate it until the tapped hole on the splice is aligned with the $\frac{1}{2}$" hole in the second pipe. Insert the second set screw and tighten both it and the first set screw so as to expand the splice and complete the union. If it becomes necessary to remove the splice, simply remove both set screws, re-compress the splice by driving the upper plate onto the lower with a punch, and pull the two pipes apart.

For individuals who would prefer not to fabricate their own splices, IESs are available in standard sizes from Art Drapery Studios, Inc., a Chicago rigging company. For more information on the Internal Expansion Splice and other creative rigging solutions, contact

Art Drapery Studios, Inc.   Chanco, Ltd.
5520 West Touhy Avenue, Unit M   3131 West Grand Avenue
Skokie, IL 60077   Chicago, IL 60622
(312) 736-9700   (312) 638-0363

When a designer asks for the seemingly impossible, often the simple low-tech solutions are the best. Having to make a drop disappear in a five-second blackout with no one onstage and no flyspace sounded impossible at first. But, with about $8.00 worth of "bulldog clips" like the one pictured in Figure 1, the drop tripped easily and predictably. We attached a series of the clips to our batten and grabbed the drop's webbing with the clips. The drop could then be pulled from the clips' grasp and dropped to the floor. The fallen drop became part of the scenery for the end of our show, but we could easily have pulled it offstage and into a waiting hamper.

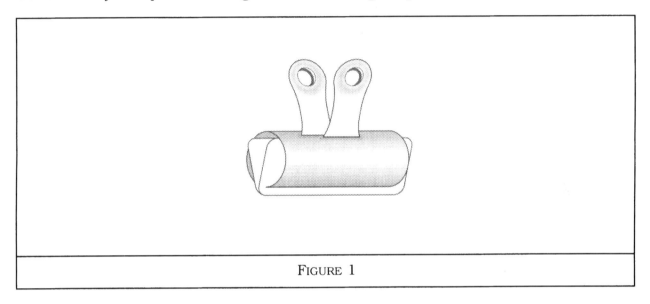

FIGURE 1

The use of bulldog clips is straightforward and easily adaptable to different situations. Heavyweight muslin or standard 3" webbing reinforcement at the top of the goods is essential, for repeated use will tear the cloth otherwise. Tie the clips onto the batten with tie line through the hole in only one of each clip's jaws. If both jaws are tied it is very difficult to get the clip to open and close correctly. To trip the drop simply pull on one end. Gravity will continue the drop's fall from one end of the pipe to the other in a controlled manner. It would be equally easy to cause the drop to fall from the ends to the center merely by tripping both ends at the same time. The goods will fall into the same place repeatedly if the operator is careful not to breast the pipe while tripping the drop.

Predictably, the heavier the drop, the more clips are needed. A 40' x 10' erosion-cloth drop used 80 clips, while a 26' x 8' chiffon drop needed only 18. The best way to determine the number of clips needed for a particular application is by trial and error. Finally, the distinctive and, in some situations, objectionable metallic clicking noise made by the clips as they snap closed after releasing the drop can be reduced by covering the inside jaws of the clips with gaffers' tape.

Phenolic Resin Pulleys: Out of the Skies and into the Future

Corky Boyd

Designed for use in aircraft, high-pressure-laminated phenolic resin pulleys are also useful to the theatre technician desiring durable, cost-effective, and dependable aircraft cable pulleys. Pulleys of the size commonly required for stage rigging can be purchased for $6.00 to $15.00 and can be reused and adapted to fill a variety of rigging needs.

The pulleys are formed when various fibers are impregnated with high-strength resins and subjected to high pressures while curing. Resulting phenolic laminates are machined into pulleys and fitted with bearings. Available in an assortment of diameters and cable sizes, the pulleys cause virtually no cable wear and suffer only minimal wear themselves when subjected to normal loads. The pulleys are also available with a variety of prefitted metallic bearings that assure a smooth and quiet operation superior to that of unrated and noisy hardware store pulleys. The bearings are manufactured in a variety of shaft diameters but are readily available in $\frac{1}{4}$" and $\frac{3}{8}$" diameters, so standard bolts may be used as axles. In critical applications, using high strength hardware may be necessary to realize the maximum safe working load of the pulley itself.

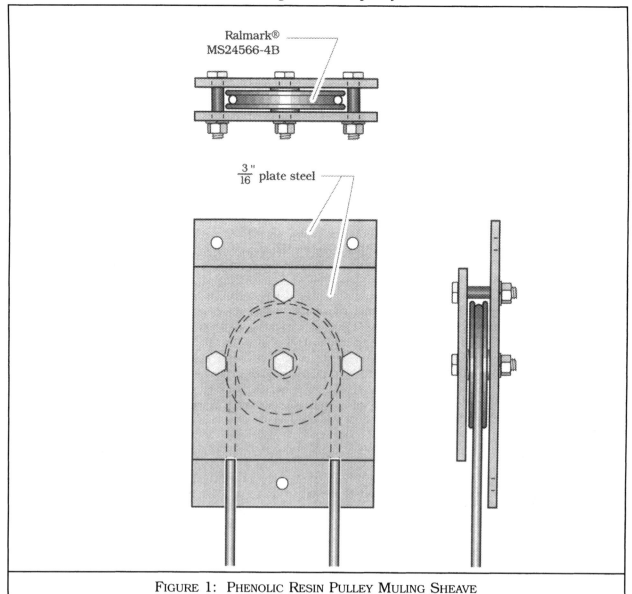

Ralmark®
MS24566-4B

$\frac{3}{16}$" plate steel

FIGURE 1: PHENOLIC RESIN PULLEY MULING SHEAVE

Since phenolic resin pulleys are designed to exacting standards, they are well suited to use in systems with critical load considerations. The deep flanges on these pulleys help to keep the cable riding securely in the groove, and manufacturer specifications are available detailing allowable load limits on pulleys and maximum cable loads for the different sizes of pulleys. Figure 1 shows an example of a shop-built sheave that uses a $3\frac{1}{2}$" pulley (Ralmark® #MS24566-4B) sized for $\frac{3}{16}$" aircraft cable sandwiched tightly between two pieces of $\frac{3}{16}$" plate steel to form a general purpose muling or turn-around sheave. The additional bolts around the perimeter of the pulley serve as stops to keep the cable from slipping out of the sheave when tension is released. At a cost of about $8.00 (less in large quantities), this pulley is rated for a working load of 1200 pounds.

If, as in the illustration, the wrap angle for the sheave is a full 180°, then the relatively small diameter of the pulley will slightly degrade the safe working load limit and service life of the cable. A larger, more expensive pulley, like the $4\frac{1}{2}$" Ralmark® #MS24566-5B (about $14.00 and rated for a working load limit of 3000 pounds), will provide an adequate diameter to preserve the maximum safe working load for $\frac{3}{16}$" cable.

Of course, designing a system requires that particular care be taken to insure adequate safety factors whenever a failure of the system or its components could pose a life-safety threat. The slight degradation of cable strength and service life resulting from an undersized sheave is often acceptable in a tracked stage effect but may be unacceptable in a scenic effect that is rigged overhead or used to move people. Specific questions about pulley performance can best be answered by a company representative, and consulting wire rope manuals can provide recommended working loads and sheave diameters for specific cable sizes.

Specifications and ordering information for phenolic resin pulleys are available from these companies:

Ralmark Company
PO Box 1507
Kingston, PA 18704
(717) 288-9331
FAX (717) 288-0902

Arvan, Inc.
1520 West 139th St.
Gardena, CA 90249
(213) 770-3700
FAX (213) 324-6634

TECHNICAL BRIEF

Rigging Techniques

Flying Drops in Limited Space

Bruce W. Bacon

Figure 1 illustrates a drop roller that operates much like a window shade and can be used to fly a full-stage drop in 18" of masked fly space. The system's advantages over conventional "olio" drop rollers are that control is achieved using a single rope, and both the roller and control ropes remain masked at the top when the drop is in its playing position. Although this procedure assumes the theater is equipped with a fly system, the roller can be adapted for use where rigged battens are not available.

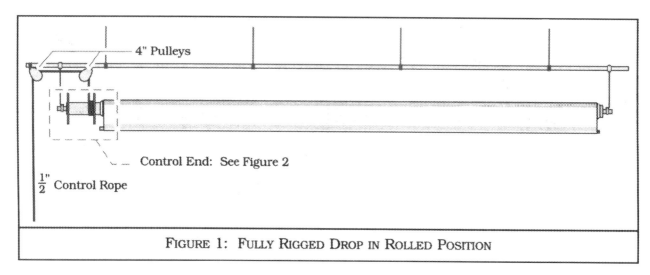

4" Pulleys

Control End: See Figure 2

$\frac{1}{2}$" Control Rope

FIGURE 1: FULLY RIGGED DROP IN ROLLED POSITION

MATERIALS

One length of 5" schedule 10 aluminum pipe, 2'-0" longer than the width of the drop.

Two lengths of $1\frac{1}{4}$" schedule 40 black pipe, approximately 2'-8" long.

Four $\frac{5}{4}$ pine washers. The inside hole should accept the $1\frac{1}{4}$" black pipe snugly, and the outer diameter should match the inside diameter of the aluminum pipe.

Two heavy-duty bearings to accept $1\frac{1}{4}$" schedule 40 black pipe.

Two C-clamps, lighting type. Each clamp must connect to a bearing as described in Step 5 of the Procedure. Obtain suitable hardware for the bearings you are using.

One length of $\frac{1}{2}$" control rope. The length should be twice the drop height plus the horizontal distance from the control position to the control end of the pipe.

Two $\frac{5}{8}$" plywood washers to channel the control rope. The inner hole should accept the aluminum pipe snugly; the outer diameter should be about 4" larger.

Two 4" pulleys to accept the control rope.

ASSEMBLY (REFER TO ILLUSTRATIONS)

1. Attach $\frac{5}{4}$ pine washers to the $1\frac{1}{4}$" black pipe shafts, making axles for either end of the roller. Each pipe has two washers, one centered $1\frac{1}{2}$" from the onstage end, the other $5\frac{1}{2}$" from the offstage end. Washers can be fixed in place with a weld bead on either side of the washer.

2. Insert $1\frac{1}{4}$" pipe axles into the aluminum pipe with the black pipe ends protruding about 4". Attach by screwing through the aluminum pipe into the $\frac{5}{4}$ pine washers.

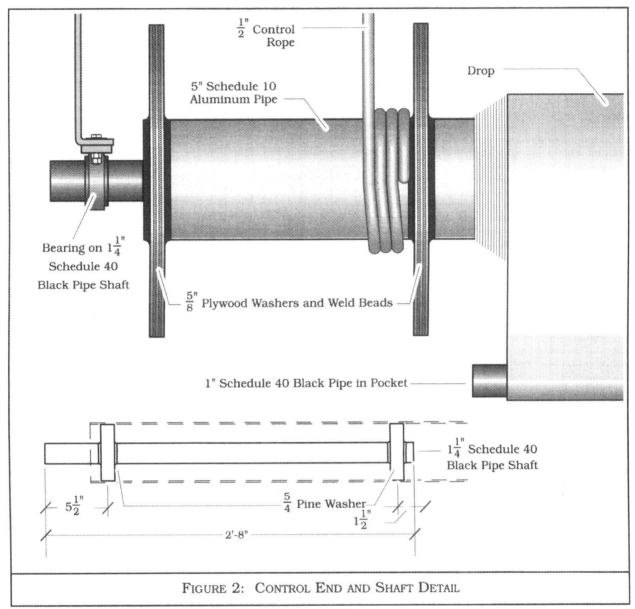

FIGURE 2: CONTROL END AND SHAFT DETAIL

3.   Attach $\frac{5}{8}$"-plywood washers to the control end of the aluminum pipe. One goes as close as possible to the end, the other about a foot away from the end. Fix in place either with weld beads or by wrapping the aluminum pipe with enough gaffers' tape on either side of the washers to prevent them from sliding.

4.   Attach the bearings to the protruding ends of the $1\frac{1}{4}$" pipes.

5.   Attach the C-clamps to the bearings. There should be about 16" between the center of the clamp jaw and the centerline of the bearing. Connections must be strong as they will support the entire weight of the system.

6.   Clamp the roller unit to the batten.

7.   Tape the top edge of the drop to the aluminum roller. Weight the drop with a 1" schedule 40 black pipe in a pocket.

8. Attach the control rope. The onstage end should be fastened securely to the roller and two wraps should be taken with the drop in the fully rolled position. Fasten one 4" pulley to the batten above the channel formed by the $\frac{5}{8}$"-plywood washers. Fasten the second pulley at the offstage end of the batten. Run the control rope through the pulleys to the offstage control position.

9. Fly the system out, keeping tension on the control rope. Tie off the rope at the control position.

OPERATION

To fly the drop in, ease the tension on the control rope. The weight of the 1" pipe in the drop's pocket causes the drop to unroll, spinning the roller. This winds the control rope onto the roller between the two plywood washers. Pulling the control rope spins the roller in the opposite direction, winding the drop back up.

NOTES

1. Aluminum is used for the roller because of its light weight.

2. The above system — a 25' roller with a 20' high drop — weighs about 150 pounds and will deflect less than $\frac{3}{4}$" in the middle. Steel is cheaper and stronger, but would be nearly three times heavier.

3. Longer rollers can be made by welding two tubes together. They, however, should be supported internally against deflection.

4. When sewing and painting a drop, allow two or three feet extra at the top for wraps on the roller.

5. New drops painted with dyes, dry pigments, or casein work best. Old paint will tend to crack.

6. The rope's bunching on the roller will not affect the smooth operation of the system.

SOURCE

The Goodspeed Opera House, East Haddam, Connecticut.

ﭏﭏﭏ

The use of wire rope and related hardware is a time-consuming rigging process. Wire rope is difficult to work with because of its tendency to kink and unravel at the ends. Applying Nicopress® sleeves or cable clamps requires the careful use of special tools, and assembly is often either slow or faulty or both.

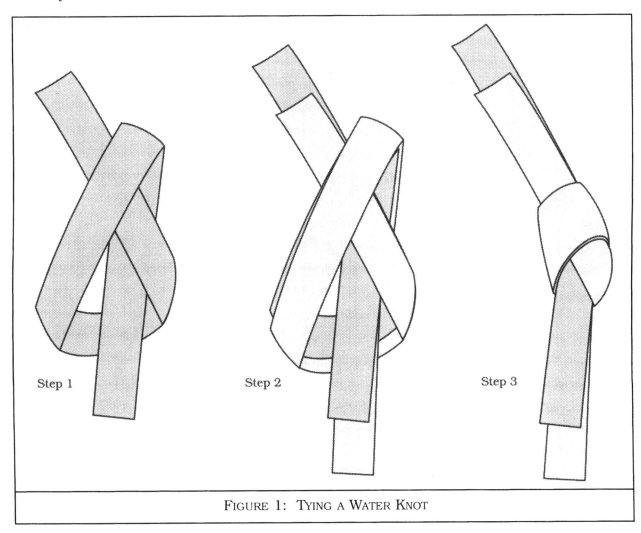

Step 1   Step 2   Step 3

FIGURE 1: TYING A WATER KNOT

A simpler and more flexible rigging system involves the use of webbing slings and carabiners. This system requires no tools, is adjustable and reusable, and makes quick connections possible. The carabiners' wide opening and their self-closing gates allow one-handed operation. The only special skill required to use this system is the ability to tie a water knot and form a choker hitch. See Figures 1 and 2.

To form a water knot, begin by tying an overhand knot in one end of the webbing. See Figure 1. Feed the other end of the webbing through the loops of the overhand knot and make it lie flat against the loop's contours. Once the overhand knot has been pulled tight, the water knot is complete.

The completed sling is attached to structural support by use of the choker hitch, and then carabiners are clipped into the sling as in Figure 2, forming a pickup from which scenery can be hung.

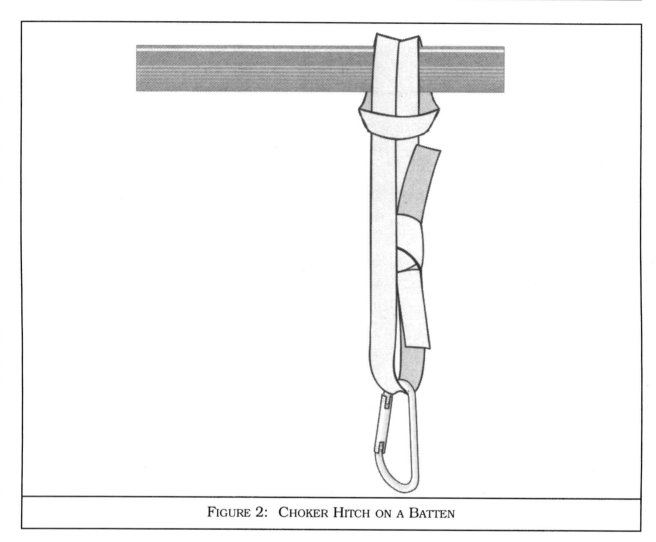

FIGURE 2: CHOKER HITCH ON A BATTEN

Obviously, slings can be attached to any support that they can be looped around, and are very useful in the quick rigging of, say, that awkward piece of masking that has to be added after all. Attaching such a sling and clipping in a carabiner takes only seconds, and when one is rigging many pickups, the use of this system can save considerable time.

The only note of caution in using this system is that, as always, it is important to check that the materials used are rated. This is especially true of webbing, as it comes in rated and unrated varieties. Suppliers of webbing and carabiners will be able to provide load limits for rated climbing gear.

There are many manufacturers of webbing and carabiners both for industrial as well as climbing/rescue purposes. Load ratings will vary from manufacturer to manufacturer, so it is important to obtain the load ratings for the specific equipment being purchased.

In general, carabiners can be purchased with load ratings ranging between 4500 and 7000 pounds breaking strength with the gate closed. Moreover, these come with standard (straight), bent, and locking gates — straight and locking gates being the most useful for theatre rigging purposes. Bent-gate carabiners are designed specifically for sport climbing applications and are not recommended for standard rigging applications.

Webbing slings, also called "runners," made with standard rated 1" tubular webbing and tied with a water knot, have a tested straight-pull breaking strength of greater than 5200 pounds. This rating assumes the raw webbing has a straight-pull breaking strength of over 3300 pounds and that the knot is tied cleanly and properly.

Any knot tied in a sling decreases the sling's effective breaking strength. A lark's head knot, for example, reduces a sling's strength by 22%, and thus, a tied, 1" tubular webbing sling hung with a lark's head knot has a breaking strength of just over 4000 pounds: 5200# x 0.78 = 4056#. There are two ways to raise breaking-strength values: hang a sling differently, or use a stronger sling.

The first method can be accomplished by simply draping the sling over the batten and clipping in to both ends. See Figure 1A. This method effectively doubles the load rating as it uses double the amount of webbing per pickup point. It is important to remember, however, that the carabiner's load rating has not changed, so that either a second one must be added, or its load rating must be used as the load rating for the system. This draping, or "basketing," of a sling will allow the pickup to slide along a batten. Adding a wrap around the batten will alleviate this problem while not significantly decreasing the pickup's load rating. See Figure 1B.

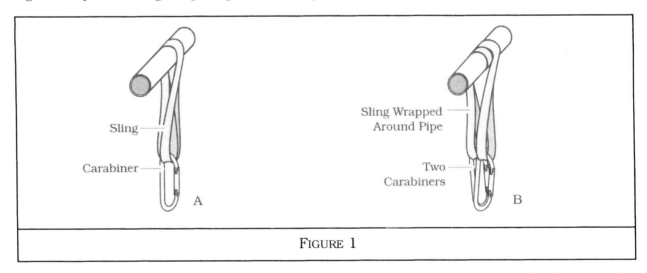

Sling

Carabiner

A

Sling Wrapped
Around Pipe

Two
Carabiners

B

FIGURE 1

Alternatively, webbing vendors or manufacturers can supply stronger slings or make them to order. There are many types of rated webbing, different widths, weaves, fibers, etc. In discussing needs with them, remember that working loads will be many times smaller than the rated strength of a piece of rigging equipment, depending on the safety factor one chooses to use.

This article describes a simple rigging system that allows both vertical and horizontal movement of scenery in two simultaneous operations. The system is useful in situations where it is necessary to fly units diagonally through space, and can also be employed to keep the backstage floor area clear of rigid set pieces that are too tall to be flown vertically out of sight. In addition, *a vista* scene shifts that take advantage of the system's potential for diagonal movement can help create very successful "cinematic" visual effects. Figure 1 illustrates this last possibility and describes the rig in general.

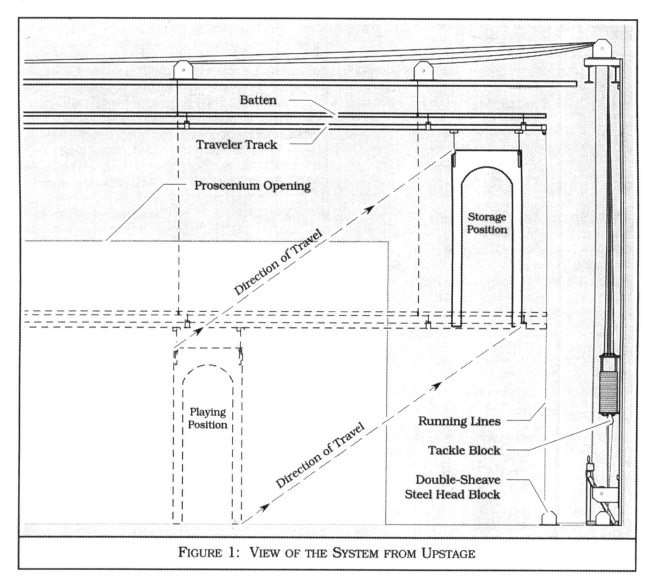

FIGURE 1: VIEW OF THE SYSTEM FROM UPSTAGE

RIGGING

The travel-fly rig uses a modified traveler track attached to a single-purchase system batten. Unusually long traveler track running lines are reeved through two floor-mounted double sheaves — one directly below the end of the track, and the other as close as possible to the lockrail — and through a single tackle block attached to the bottom of the lineset arbor. This arrangement keeps tension on the running lines constant as the batten and arbor move. See Figure 2 for details.

OPERATION

Proper operation of the system requires two people. The flyperson at the lockrail controls vertical movement, while the travel operator at the floor-mounted sheaves below the end of the track controls horizontal movement.

Consistent and smooth operation of this system is possible only with thorough rehearsal. Since the speed at which the travel operator works depends on the direction and speed of the system's vertical movement, coordination between both operators is critical.

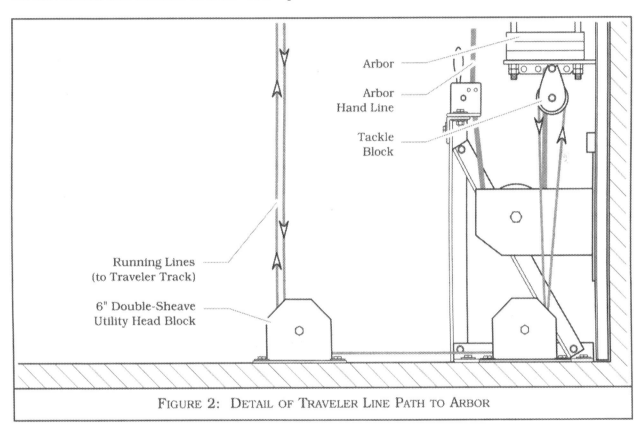

Arbor

Arbor
Hand Line

Tackle
Block

Running Lines
(to Traveler Track)

6" Double-Sheave
Utility Head Block

FIGURE 2: DETAIL OF TRAVELER LINE PATH TO ARBOR

SAFETY NOTES

Traveler running lines crossing the backstage floor present a serious tripping hazard. Mark and light the area involved or build a deck over the lines. Also, provide firmly anchored stops at the ends of the traveler track.

After reading Don Holder's "A Simultaneous Travel-Fly Rig," Paul Carter wrote the editors, "We just hung a show, *Brigadoon*, here at the [New York] State Theatre that used a similar (travel-fly) system . . . and avoided all the rope on the stage floor and rigging to the arbor.

As the traveler moves up and down with the pipe, the system works much like [the original]. In addition, this system works when the fly rail is not at stage level."

The modified system is illustrated in Figure 1.

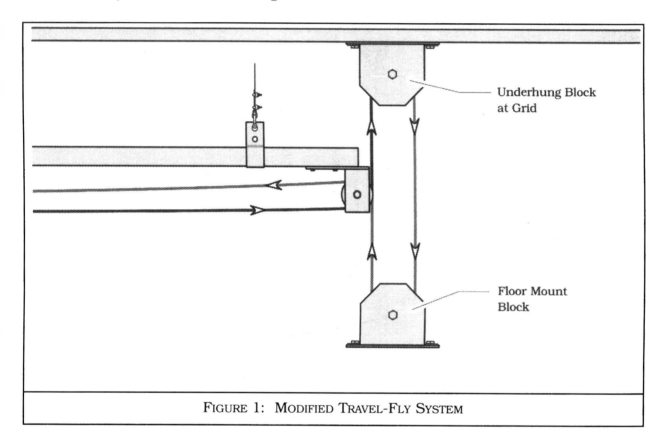

FIGURE 1: MODIFIED TRAVEL-FLY SYSTEM

Wire-rope curtain tracks are not new. However, such systems are often noisy and can be quite bulky. We developed the following system to meet the following requirements:

1. to support a continuous bobbinet panel 60' wide, 11' high,

2. to incorporate a very slim, visually unobtrusive, curtain track,

3. to allow movement quiet enough to hear dialogue over,

4. to allow for smooth movement of the curtain by an actor onstage,

5. to restrict maximum sag to 6" at the center of the track.

MATERIALS

A 20-yard length of bobbinet created a curtain panel weighing 8 pounds. White twill tape, with $\frac{1}{8}$" grommets every foot, held $\frac{1}{2}$" split-key rings as glides and acted as a top web. The track itself was $\frac{1}{8}$" wire rope with Nicopress® swages or wire rope clips forming a thimble loop on one end of the line, where it was anchored to the wall. In place of a thimble, a marine wire pulley was used on the other end in order to allow the wire rope to slide through the wire rope clips during the initial tightening. A #1-size polyethylene tube was used over the wire rope to give the rings a smooth surface over which to slide. This polyethylene tube had an inside diameter of 0.17" and an outside diameter of 0.25".

The tubing cost $21.45 for a 500' spool, and is available from most plastics distributors. The wire rope and other rigging components cost about $30.00.

INSTALLATION OF TRACK AND CURTAIN

1. Cut the wire rope to 70' to allow for tightening of the loop.

2. Cut the polyethylene tube to 60' length.

3. Lubricate the wire with silicone or WD-40® and push the wire through the tubing laid out in a straight line.

4. Feed the wire rope through the rings of the finished curtain.

5. Connect the thimble end of the wire rope to the anchoring point.

6. Feed the wire rope through the pulley and make the tightening end loop.

7. Use the end loop to pull the wire rope taut with a winch, and trim the excess tube.

8. Tighten the clips at the pulley connection.

9. Lubricate the outside of the polyethylene tube with silicone spray so that the curtain runs quietly and smoothly.

SAFETY CONSIDERATIONS

This curtain track contains a great amount of potential energy, and must be handled with a great deal of care. The tension in the wire rope with the curtain suspended must be calculated and rigging components selected with an appropriate margin of safety.

For a production of *As You Like It* at Mount Holyoke College, the technical staff was asked to fly and trip a scrim rapidly and in the dark. Clearly, life would be simpler if the movement of the counterweight arbor could be made to trip the bottom of the scrim at twice the speed of the top of the scrim. The double purchase principle came to mind with the thought of a 2:1 ratio.

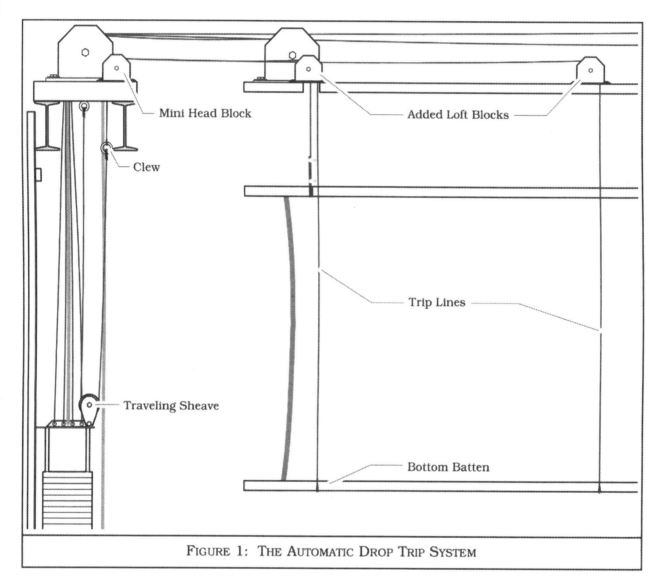

FIGURE 1: THE AUTOMATIC DROP TRIP SYSTEM

Deanna Chrislip, Technical Director for the show, put a lightweight one-inch-square wooden batten in the bottom of the scrim and tied four triplines (mason line) to it. The triplines went straight up to add-on loft blocks in the grid, over to a mini head block, and down to a clew (they all joined at a ring). From the clew a $\frac{3}{8}$" line ran down to a pulley mounted on the top of the scrim's counterweight arbor and then back up to the grid. See Figure 1. In this manner as the arbor moved down one foot, drawing the top of the scrim up one foot, the trip lines were pulled up two feet. This system works best on light drops.

❧❧❧❧❧

Fying flats that, at low trim, are not parallel to the plaster line can be a challenge. The rig described here provides a means to fly a flat on a single lineset and have it rotate into a different orientation automatically as it flies in or out.

OPERATION

In this rig, four lines are attached to each of two pickups on the flat. See Figure 1. Lines **a** and **b** are attached to a batten; lines **c** and **d** pass through pulleys on the batten and are secured at the grid directly above the batten pickups. At low trim, all four lines share the load, and line tension holds the flat at an angle to plaster. During the out, **a** and **b** gradually assume the full load, swinging the flat parallel to plaster.

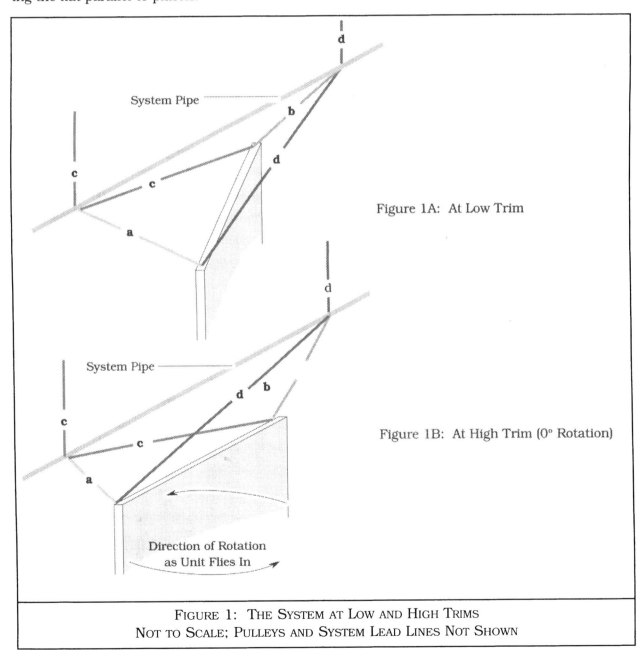

Figure 1A: At Low Trim

Figure 1B: At High Trim (0° Rotation)

Direction of Rotation as Unit Flies In

FIGURE 1: THE SYSTEM AT LOW AND HIGH TRIMS
NOT TO SCALE; PULLEYS AND SYSTEM LEAD LINES NOT SHOWN

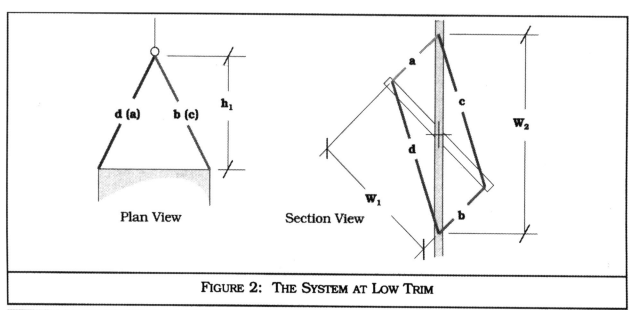

FIGURE 2: THE SYSTEM AT LOW TRIM

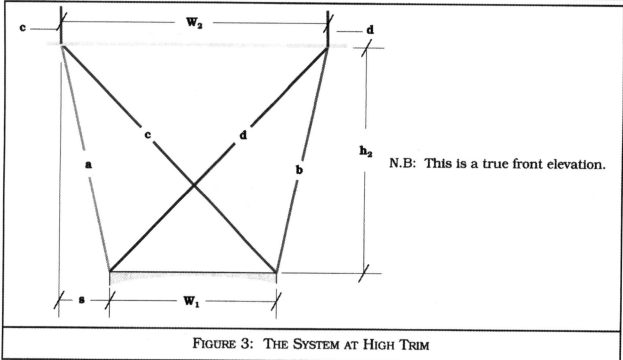

FIGURE 3: THE SYSTEM AT HIGH TRIM

DESIGN CONSIDERATIONS

Rigs like these must meet well-defined structural requirements. In ordinary flat-to-batten rigs, both the batten and the top rail of the flat behave like beams. In this design, on the other hand, the flat's rotation adds a compressive force to both members, causing them to behave like columns, which must meet the demands of slenderness ratios. These ratios would limit the pickup-to-pickup width of a soft-covered, traditionally framed flat's 1x3 top rail to a maximum of $37\frac{1}{2}$". Similarly, the pickup-to-pickup distance of a standard $1\frac{1}{2}$" schedule 40 black pipe batten would be limited to about 10'. Careful structural analysis should be applied before committing to the use of any such rig.

Having determined that the structural requirements can be met, the designer of the rig will ordinarily need to determine whether the flat can be masked at high trim. The answer to that question can be determined by applying some fairly simple formulas to data obtained from the set designer's drawings.

In Figure 2, h_1 represents the minimum distance required between the top rail of the flat and the system pipe in order to mask the batten at low trim. In Figure 3, w_1 represents the pickup-to-pickup width of the flat; and w_2 the pickup-to-pickup distance along the batten. The difference between w_2 and w_1 should be made as great as possible in order to minimize the momentum of the flat's rotation as it is brought parallel to plaster on the way out. Using a compass (as illustrated in Figure 4) and a scale, the designer can discover s, the horizontal distance between flat and pipe pickups, and can scale m, the horizontal measure of line a. The true length of line a is determined by using the following formula:

$$a = \sqrt{h_1{}^2 + m^2}$$

Having defined a and s, the designer can calculate h_2, the distance between the top rail and the batten by using a second formula:

$$h_2 = \sqrt{a^2 - s^2}$$

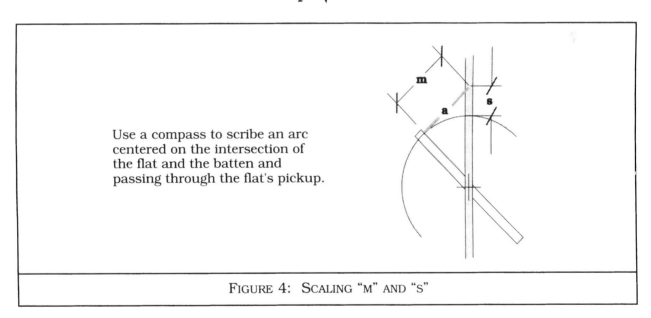

Use a compass to scribe an arc centered on the intersection of the flat and the batten and passing through the flat's pickup.

FIGURE 4: SCALING "M" AND "S"

Knowing h_2 (and the flat's vertical dimension) the designer can sketch the flat on the general section and discover whether or not high-trim masking is possible.

FINAL NOTES

This rig requires further careful planning. Before committing to its use the technical designer will have to address three remaining issues: the inter-batten clearance required by the lines during rotation; the size of the wire rope to be used in order to accommodate the rotation-induced increase in tension; and the change of weight that occurs as the rig's weight transfers from one pair of lines to another. Nevertheless, when such a rig can be used, it produces a stunning visual effect.

With increased concern over fleet angles, the Mount Holyoke technical staff have been less inclined to give in when designers ask to have loft blocks moved "just a few inches upstage." In order to meet designers' needs, we've begun using C-clamps as shown in Figure 1 to offset the placement of very light scenic units without having to move loft blocks. In such rigs, a system pipe is prevented from turning by clamping one or two side arms to the pipe and attaching them to the pipe's lead lines, as is often done when electrics are overhung, rather than hung directly under their pipe. We usually side-arm the lines that are carrying most of the unit's weight. Once the pipe has been rigged to prevent rotation, the unit's pickups are hung from the system pipe and then passed over C-clamps mounted horizontally on the pipe as shown in the illustration.

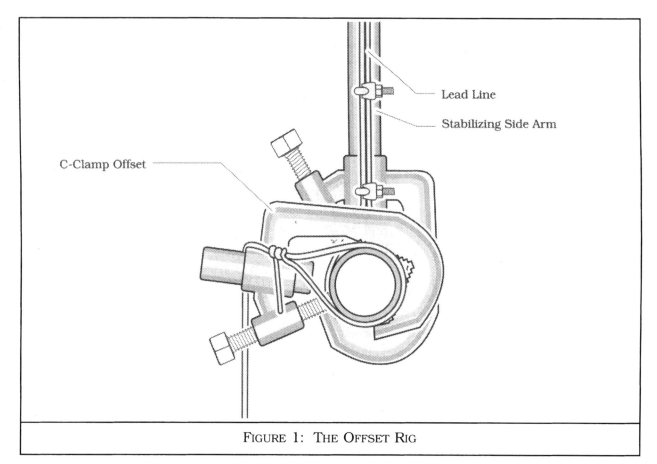

C-Clamp Offset

Lead Line

Stabilizing Side Arm

FIGURE 1: THE OFFSET RIG

As might be expected, this approach has some important limitations: it should not be used to offest units weighing more than about 20 pounds; nor should it be used to achieve offsets of more than 3" or so. And even within these limits, some frustrations will occur. If you want to hang a unit totally upstage or downstage of the pipe, the side arm's C-clamp will tend to swing in the opposite direction, returning the unit to the plane of the pipe's lead lines. Some of this effect can be minimized by the use of longer side arms, however, and a 10- to 20-pound unit can be moved upstage or downstage about 3". On the other hand, this rig works particularly well if you want the unit hung with one end downstage of the pipe and one end upstage of the pipe, for then the unit's weight is centered under the pipe though not parallel to the pipe. Nevertheless, the side arms are still needed to hold the pipe steady while adjusting the trim.

For a production of *Peter Pan* I designed a flying system to enable an actor to fly across the stage, parallel to the proscenium, with independent control of both his height above the stage and his lateral position. The system, described here and illustrated in Figure 1, was simple to operate, reliable, and safe.

Two 30' pieces of heavy-duty steel channel (#2800) were spliced together to provide a 60' track. This track was hung from a batten using hanging clamps placed on 4' centers. Tormentors were set to reduce the proscenium width to 40', thus giving us 10' of track offstage left and right.

Transverse movement was accomplished by a tram equipped with three master carriers (#2852) that allowed it to travel along the track. The 4' tram carried two underhung traveling blocks that would be used for the vertical movement of the actor. A live-end pulley (#2863-A), dead-end pulley (#2864-A), and floor pulley (#2866-A) were used to rig the tram in a standard traveling fashion.

A 2:1 mechanical advantage was designed into the system for the vertical movement of the actor. A $\frac{1}{8}$" aircraft cable connected to a 1" operating line ran through an underhung head block mounted on the track, through the first traveling block on the tram, down and around a floating block, up and through the second traveling block, and finally, to an end stop on the far end of the track. The actor was attached to the floating block by appropriately sized black aircraft cable.

As the tram moved along the track, the $\frac{1}{8}$" aircraft cable rode through the two traveling blocks and the floating block without changing length. Thus, unless the 1" operating line was raised or lowered, the actor remained at a constant height as he was flown across the stage. When the 1" operating line was pulled down, the floating block rose, lifting the actor half of that distance. The rig used two operators: one to control vertical movement, the other to control horizontal movement.

The trim height of the batten holding the track was determined by adding the height of the set's vertical sight lines to the maximum height of the flights and increasing that sum by an additional 10'. The trim height of the batten for our production, 45' above the stage floor, helped reduce the noise made by the tram's travel along the track. Since the arbor was counterweighted only for the weight of the rigging, the system became extremely pipe-heavy when an actor was flown. To keep the rope lock from slipping and the batten from dropping during flights, one end of a piece of $\frac{1}{8}$" aircraft cable was secured to the bottom of the arbor and the other to the locking rail of the counterweight system.

All the blocks used in this system had 8" sheaves with Timken Roller Bearings to help ensure smooth running. The cost for the flying system described here was less than $1,100.00.

The system ran flawlessly throughout rehearsals and run. The operators found the system easy to operate, and everyone involved in the production felt secure in the safety of the flying. This system met all of the flying requirements of the script and allowed us to choreograph spectacular sequences.

SUPPLIERS

Blocks were supplied and the tram built by SECOA, 2731 Nevada Avenue North, Minneapolis, MN 55427. Track and hardware were built by Automatic Devices Company (a SECOA dealer), 2121 South Twelfth Street, Allentown, PA 18103. Black cable can be obtained from Foy Inventerprises, 3275 East Patrick Lane, Las Vegas, NV 89120; and black-vinyl-coated cable, from Hartford Cordage and Twine Company, 687 Cedar Street, Newington, CT 06111.

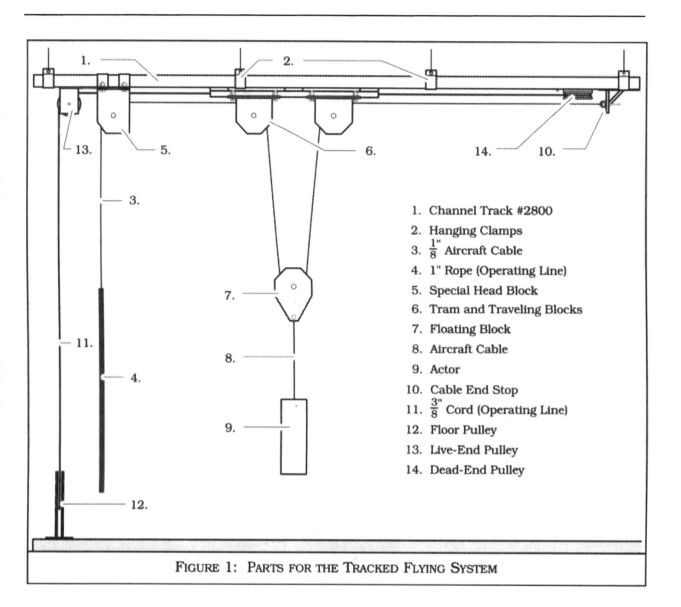

1. Channel Track #2800
2. Hanging Clamps
3. $\frac{1}{8}"$ Aircraft Cable
4. 1" Rope (Operating Line)
5. Special Head Block
6. Tram and Traveling Blocks
7. Floating Block
8. Aircraft Cable
9. Actor
10. Cable End Stop
11. $\frac{3}{8}"$ Cord (Operating Line)
12. Floor Pulley
13. Live-End Pulley
14. Dead-End Pulley

FIGURE 1: PARTS FOR THE TRACKED FLYING SYSTEM

Structures that appear to defy the basic laws of physics and gravity are part of the magic of theatre. One such structure is the apparently sag-free catenary cable curtain track that stretches across the entire expanse of a stage and disappears into the wings left and right. For no matter how tightly a cable is stretched through mid-air, it can never be entirely sag-free, whether it supports any other object or not.

The most common way to provide a sag-free cable curtain track is to apply as much tension as possible with a come-along or turnbuckle. Tremendous forces are generated in such systems, and to protect the actors in a Yale Repertory Theatre production I felt it was important to limit the pulling strength of whatever tensioner we used. Two cables stretched across the 56'-wide stage needed to carry four 8' x 10' panels of chain-weighted ripstop nylon. I wanted to insure that, if an actor were to trip and grab the curtain, the system's tensioner would release momentarily. Further, the production was in rotating repertory, and the system would have to be set and struck repeatedly. Figure 1 illustrates the system I designed to meet these criteria.

In order to calculate the amount of tension needed to pull the cable taut, you must first make an assumption about how much sag will be acceptable. In my case, the scenic designer agreed that 3" of sag in a 56' span would be acceptable. I calculated that if the cable could sag no more than 3" when the full weight of the curtains was bunched in the center of the span, it would be subjected to 790 pounds of tension. The $\frac{3}{16}$" vinyl-coated cable I had planned to use had a breaking strength of 3700 pounds and was easily capable of carrying this load with a 5:1 safety factor.

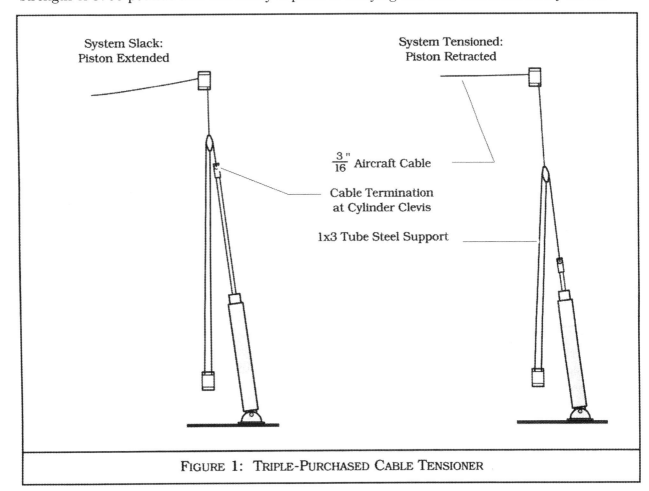

System Slack:
Piston Extended

System Tensioned:
Piston Retracted

$\frac{3}{16}$" Aircraft Cable

Cable Termination
at Cylinder Clevis

1x3 Tube Steel Support

FIGURE 1: TRIPLE-PURCHASED CABLE TENSIONER

Recognizing the inherent compressibility of air, I decided that a pneumatic cylinder would be a good choice for a tensioning device. The downstroke of a 7" or 8" pneumatic cylinder could easily take up an extra 6" of cable, which would be enough to allow a member of the shift crew to disconnect the cable every night at changeover. However, since a cylinder capable of creating 790 pounds of tension with a 6" throw was rather large and expensive, I decided to triple-purchase the cable, making use of the pneumatic cylinders available in our stock. The triple-purchased system used an 18"-throw cylinder with a $2\frac{1}{2}$" bore. Its 3:1 mechanical advantage created the necessary force, and the cylinder's 18" throw reduced 6" of slack to an acceptable 3" sag.

The system worked well: extra weight applied to the cable simply resulted in the further extension of the piston rod rather than greater tension in the cable. The sag in the center of the curtain proved to be very close to the desired 3", and the changeover crew was able to install and strike the cables safely in less than a minute.

❧❧❧

Using the principle that tracks drafting-board parallel rules, Richard Gold, Technical Director for the production of *Mississippi Nude* at the Yale School of Drama, enabled a single operator to move a curtain track upstage and downstage. The movable track was hung parallel to the plaster line and perpendicular to the three stationary tracks from which it was suspended. Aircraft cables **A** and **B**, which ran through sheaves mounted at the ends of the movable track, were secured at the ends of the outer stationary tracks, forming the same "X" rig that guides a parallel rule. Figure 1 presents the rig in plan; Figures 2 and 3, its details. Adding a running line to one end of the movable track allowed a single operator to move the track upstage and downstage with ease.

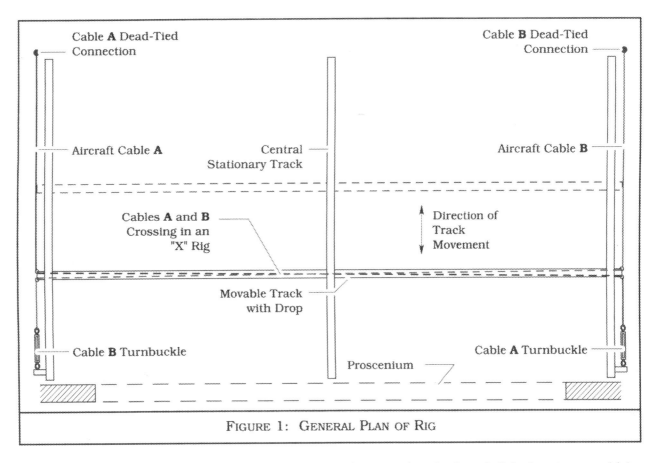

FIGURE 1: GENERAL PLAN OF RIG

This principle assures parallel movement in any plane, and with thoughtful planning, could be applied to wagons, to trap platforms, and to hard-covered traveler panels. Depending on the scale of the application, it could be necessary to devise a means of keeping the cables from sagging. Regardless of the specific application, it is always important that sheaves be secured to take forces in all directions. It is equally important to tighten the guide cables alternately, checking to make sure that the guided object's desired orientation is maintained: tightening one cable all the way first will rotate the object out of its desired orientation, and such misorientation can jam the sheaves and/or overstress other hardware.

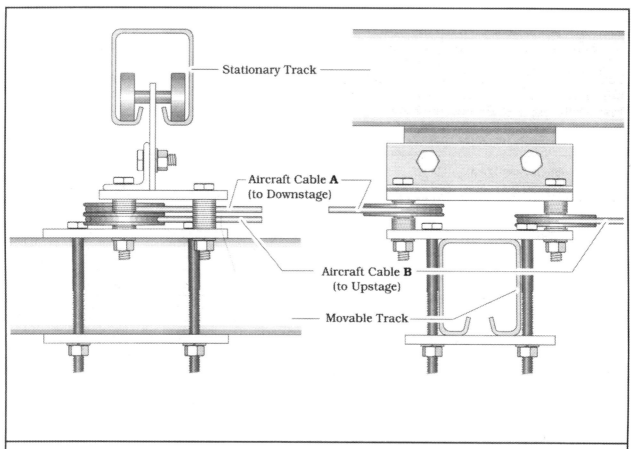

FIGURE 2: SL GUIDE ASSEMBLY SEEN FROM US AND SL

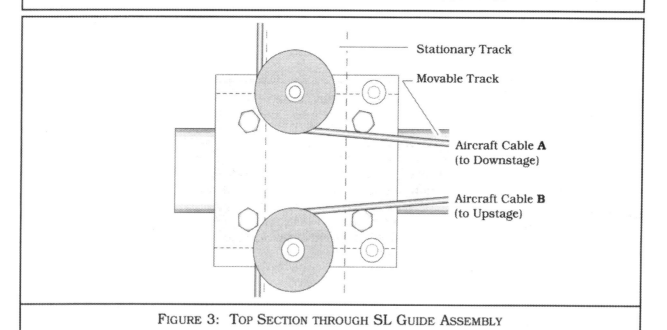

FIGURE 3: TOP SECTION THROUGH SL GUIDE ASSEMBLY

TECHNICAL BRIEF

Safety

The configuration of structures leading from one level to another on stage is highly design dependent. However, several safety considerations should be noted and, as much as possible, observed in the construction of step units and ladders.

The general workable incline range for covered step units is 20° to 50°, with the most comfortable incline about 30° to 35°, or a 7" to 9" riser with a 9" to 12" tread. Changes in riser height or tread width within a set of stairs should be avoided. If the risers or treads must vary, the difference should be less than $\frac{1}{2}$". If it is impossible to meet this criterion, a landing may be incorporated to provide a break between two sets of stairs with different riser heights and/or tread widths. If a single step must vary in height or width, it is safest to place it at the bottom of the unit.

As the riser height approaches the lower safe limit for stairs (about 3"), the depth of the tread should be increased. Riser height should never exceed 12", and as the pitch approaches the upper safe limit (about 10"), an open riser unit should be considered. If, for example, a rise of 3' for each foot of run were desired, an open riser unit might be designed with 12" risers and 9" treads overlapping 5" (*i.e.*, with 4" exposed in plan). Obviously, railings become important as the pitch increases.

If pitches steeper than 75° are desired, a rung or wall ladder is the safest solution. A 12" rise between rungs should be considered maximum. The wall ladder, however, has an additional constraint: a minimum of 7" of toe room between the ladder and the wall.

ε♠ε♠ε♠

The following notes are intended to assist the theatre technician or producer in complying with accepted practice in fire safety procedures. As state and local codes and enforcement vary widely, each producer should check with local fire safety officers and enlist their advice and cooperation to insure compliance with the code.

FABRICS

All fabrics used in scenery or scenic decoration should be either inherently flameproof synthetics or all-natural fibers that have been processed for flameproofing. The standard test is to hold a match to the bottom edge of a 4"-wide strip of fabric for 10 seconds. When the match is removed, the flame should go out though the fabric may be charred. Fabrics purchased pre-processed should be accompanied by a certificate provided by the processor. Treatment can be done easily in the shop with the use of commercially available flameproofing compounds.

WOOD

All wood and wood products used in scenery must have all exposed surfaces painted with either casein or latex-based paint or conventional water-based paint to which a flameproofing compound has been added.

PLASTICS AND STYROFOAM®

Sheet plastics and foam products pose a particular problem. Some vinyl and acetate products are available that are called "self-extinguishing" and should meet most code criteria. Foam insulating materials with low fire ratings are also available. In general, however, it is best to cover all exposed plastics and synthetics with a flameproof fabric and a coat of water-based paint.

CARPETING

Carpets either must be composed of 100% natural fabrics and then treated for flameproofing, or must have been tested and rated according to standard National Fire Protection Association tunnel tests. Retailers of industrial and institutional carpets can provide certificates of test results for lines of carpet that they sell.

OPEN FLAMES AND FIRES

Caution must be exercised at all times when smoking, lit candles, torches, or matches are to be used in performance. Proper receptacles and extinguishers must be provided in the wings or off-stage, and, prior to the opening of the show, the producers should be prepared to offer local fire safety officers a careful and specific description of the use and handling of flame or fires onstage.

FIRE EXTINGUISHERS

All stages, whether temporary or permanent, should be equipped with properly charged and dated fire extinguishers located conveniently and prominently in the stage area.

AISLES AND EXITS

Aisles and exits from auditoriums must be kept completely clear and must be lighted and marked at all times. When setting up temporary seating in halls or large rooms, careful attention must be paid to allowing sufficient aisle width. Again, local codes vary considerably, but a standard rule of thumb calls for a minimum aisle width of 3'-0", with a 4'-0" to 5'-0" provision for entrance and exit routes. In general, no seat may be more than 6 seats away from an aisle, and all risers must have safety rails. It is desirable (and sometimes required) to have the seats fastened to the floor or at least connected in rows. In locations where regular inspections by fire safety officials occur, every effort should be made to have the seating arrangement inspected prior to each first public performance.

HOUSEKEEPING

The accumulation of trash and waste materials in the stage area during construction and rehearsal periods is one of the most common fire hazards. At the close of each work period, return all tools, paints, and materials to storage and remove trash from the building.

ELECTRICITY

Electricity is the most common cause of fires today. Insure that all tools and extension cords are grounded, that lighting equipment is well insulated and properly mounted, and that all wiring and circuits are sufficient to carry the loads in use.

EDITORS' NOTE

Laws governing fire safety have changed radically since this article's original publication, and vary from one locality to the next. Consult with state and local fire authorities for information about those regulations currently in effect in your area.

ৈৱৈৱৈৱ

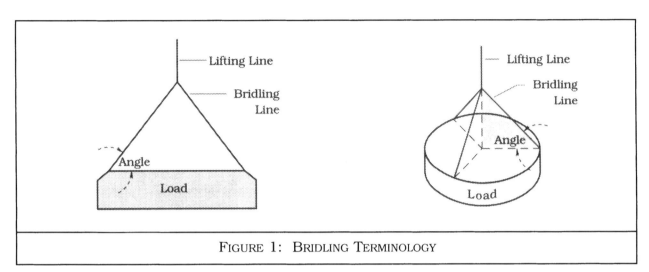

FIGURE 1: BRIDLING TERMINOLOGY

When supporting a load with two or more bridling lines, the tension in each line is greater than the load divided by the number of lines. This is true because each bridling line pulls against the other(s) in addition to supporting its share of the load. See Figure 1. As the angle between the bridling lines and horizontal decreases, the tension in the lines increases. The table in Figure 2 provides multiplying factors used to determine the tension in each of the bridling lines.

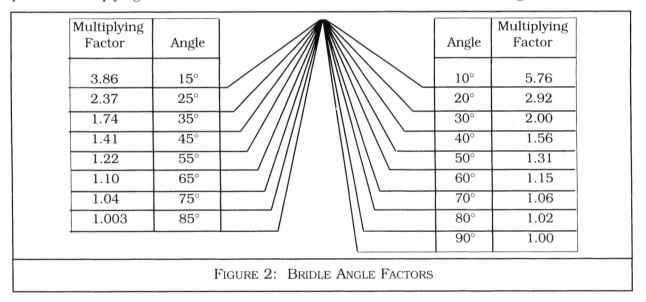

Multiplying Factor	Angle		Angle	Multiplying Factor
3.86	15°		10°	5.76
2.37	25°		20°	2.92
1.74	35°		30°	2.00
1.41	45°		40°	1.56
1.22	55°		50°	1.31
1.10	65°		60°	1.15
1.04	75°		70°	1.06
1.003	85°		80°	1.02
			90°	1.00

FIGURE 2: BRIDLE ANGLE FACTORS

To determine the tension in each bridling line, divide the load by the number of bridling lines and multiply that figure by the factor listed for the angle between the bridling line and horizontal. Example: Lifting a circular platform with three evenly spaced bridling lines (see Figure 2) where the load is 500 pounds and each bridling line is 45° from horizontal. Solution: Each line supports one-third of the 500-pound load or 166 pounds. Multiplying that figure by the factor for a 45° angle reveals that the tension in each line is 235 pounds: 166# x 1.41 = 235#.

This article does not deal with uneven loads or forces due to acceleration or deceleration. In choosing lines, always apply an appropriate safety factor.

Allowable Loads on Lumber-Nail Joints

John Marean

Figures 1 and 2 show two ways to support the non-legged platform **B** by resting it on a wood carrying strip that is nailed to an adjacent legged platform **A**. Most builders would agree that, materials and loads being equal, the carrying strip in Figure 2 could be attached to platform **A** with fewer nails than that in Figure 1. The question, however, remains: how many nails must be used in either case? This article provides a guide for answering that question.

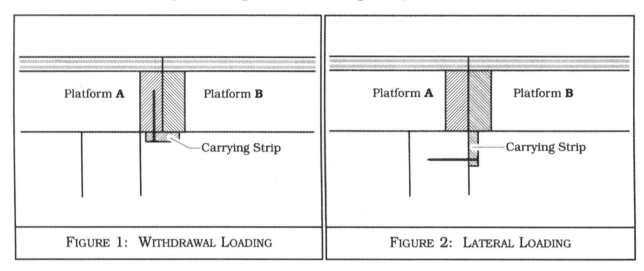

| FIGURE 1: WITHDRAWAL LOADING | FIGURE 2: LATERAL LOADING |

DEFINITION OF TERMS

Withdrawal load: A load applied parallel to a nail's shaft.
Lateral load: A load applied perpendicular to a nail's shaft.
Main member: The piece of wood in which a nail's point is buried.
Depth of penetration: The distance a nail extends into a main member.

A nail's resistance to withdrawal loading depends on its diameter, its depth of penetration, and the specific gravity of the main member. A nail's resistance to lateral loading, on the other hand, is limited by both the diameter of the nail and the wood fibers' resistance to crushing.

The formula governing withdrawal loading is:

$$P = 1380 \times G^{\frac{5}{2}} \times D$$

where:

P = resistance to withdrawal (pounds per inch of penetration)
G = the specific gravity of the main member
D = the diameter of the nail (inches, in.)
1380 = an empirically determined constant

The formula governing lateral loading is:

$$P = K \times D^{\frac{3}{2}}$$

where:

P = resistance to lateral failure (pounds per nail at rated penetration)
K = the empirically determined coefficient of density of the main member
D = the diameter of the nail (inches, in.)

Note: this formula assumes a penetration of 10 x the diameter of the nail for Group II wood species, 11 diameters for Group III, and 14 diameters for Group IV.

SAMPLE PROBLEM

The 1x3 carrying strips illustrated in Figures 1 and 2 must carry 400-pound loads. They will be attached to 2x4 Douglas fir main members by means of 6d common nails. How many nails are required in each case?

Withdrawal Loading (Figure 1)

$$P = 1380 \times G^{\frac{5}{2}} \times D$$

G (Douglas fir) = 0.51
D (6d Common) = 0.113 in.

$P = 1380 \times (0.51)^{\frac{5}{2}} \times 0.113$ in. = 28.97 pounds resistance per inch of penetration. The depth of penetration = nail length minus thickness of the carrying strip = $2" - \frac{3}{4}" = 1\frac{1}{4}"$. Thus, each nail can resist $1\frac{1}{4}"$ x 28.97 pounds per inch of penetration or 36.21 pounds. Carrying a 400-pound load would take 12 nails (400 ÷ 36.21 = 12).

Lateral Loading (Figure 2)

$$P = K \times D^{\frac{3}{2}}$$

K (Douglas fir) = 1650
D = the diameter of the nail (inches, in.)

$P = 1650 \times (0.113$ in.$)^{\frac{3}{2}} = 62.68$ pounds resistance per inch of penetration. Carrying a 400-pound load would take 7 nails (400 ÷ 62.68 = 7).

NOTES

1. Joints based on these calculations assume more than one nail is resisting the loading.

2. Nails must not split either the main member or the piece attached, and must not be too close to either member's edge or end.

3. The formulas given do not apply when subjected to impact loading such as those caused by jumping or dancing.

4. As predicted, lateral loading requires fewer nails than does withdrawal loading.

TABLES

Table I on the next page gives data for a few commonly used nails; Table II, data for a few commonly used wood species. Table III is a composite of the sort you may want to make up covering nails and wood species used in your shop. Complete tables and further information are presented in *Understanding Wood* by R. Bruce Hoadley, 1980, The Taunton Press, and *National Design Specifications: Wood Construction*, 1982 Edition, The National Forest Products Association.

Allowable Loads on Lumber-Nail Joints

John Marean

Type of Nail							
Box				Common			
6d	8d	10d	16d	6d	8d	10d	16d
Length in Inches				Length in Inches			
2.0	2.5	3.0	3.5	2.0	2.5	3.0	3.5
Diameter in Inches				Diameter in Inches			
0.099	0.113	0.128	0.135	0.113	0.131	0.148	0.162
TABLE I							

Species	G	K	Species Group
Douglas Fir	0.51	1650	II
Hemlock Fir	0.44	1350	III
Pennsylvania White Pine	0.38	1080	IV
TABLE II			

Nail Type	Douglas Fir		Hemlock Fir		PA White Pine	
	W^1	L^2	W^1	L^2	W^1	L^2
6d box	25.38	51.40	17.54	42.05	12.16	33.64
8d box	28.97	62.68	20.03	51.28	13.88	41.02
10d box	32.81	75.56	22.68	61.82	15.72	49.46
16d box	34.60	81.84	23.92	66.96	16.58	53.57
6d common	28.97	62.68	20.03	51.28	13.88	41.02
8d common	33.58	78.23	23.22	64.01	16.09	51.21
10d common	37.94	93.95	26.23	76.86	18.18	61.49
16d common	41.53	107.59	28.71	88.03	19.90	70.42

[1]Withdrawal resistance is stated in pounds per inch of penetration.
[2]Lateral resistance is stated in pounds per nail.

TABLE III

❧❧❧

When lowering flown scenery in to final trim, theatre technicians frequently need to apply braking devices to hand lines in order to achieve a safe rate of movement and to assure that movement can be stopped precisely at trim. Most technicians are familiar with the practice of "taking a wrap" around a fixed pipe, and most understand intuitively that the braking effect can be increased by increasing the number of wraps taken. But, as this article indicates, taking wraps is neither the only nor perhaps the most desirable braking device. The figures and table that follow illustrate five alternative braking devices and present the results of a series of tests performed to compare the effectiveness of those devices.

THE TEST

A 25-pound load (the brake load) was applied to each rope to be tested. The rope was then passed around a sheave, past the braking device, and around a second sheave, after which a second load was applied. The second load was gradually increased until slippage through the brake began. The ropes used in the tests were all laid ropes showing a moderate amount of wear.

THE ILLUSTRATIONS

Figure 1 illustrates both the test setup and the familiar wrap brake; Figure 2, the munter hitch, which, like the wrap, requires no additional hardware. The more complicated carabiner one-wrap in Figure 3 uses a single carabiner attached to a fixed support such as a counterweight rail-mounted sunday; and the carabiner brake in Figure 4 employs two more carabiners. Figure 5 illustrates the figure-eight brake, which is accomplished by reeving the hand line through a "figure-eight," a piece of hardware that can be purchased from mountain-climbing suppliers.

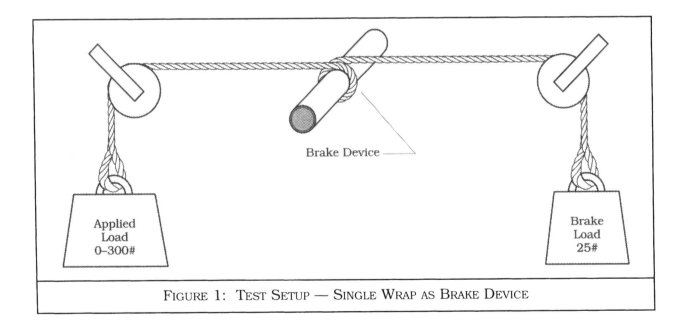

Brake Device

Applied
Load
0–300#

Brake
Load
25#

FIGURE 1: TEST SETUP — SINGLE WRAP AS BRAKE DEVICE

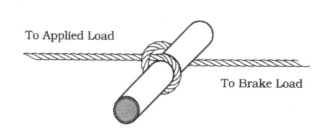

FIGURE 2: THE MUNTER HITCH

FIGURE 3: THE CARABINER

FIGURE 4: THE CARABINER BRAKE

FIGURE 5: THE FIGURE-EIGHT BRAKE

NOTES

The reader should keep in mind two important points. First, the data that follow represent the results of a small number of tests and, consequently, are more reliable as a comparative rather than definitive guide. Second, these were tests of static — not dynamic — conditions. The results indicate merely the size of the load required to induce slippage. They do not indicate the amount of braking force required to stop that slippage once it has begun; and slippage will continue once it has begun unless the braking force is increased appropriately — even if the load is reduced.

BRAKE TEST RESULTS

Brake Device (all applied to $1\frac{1}{4}$" schedule 40 black pipe)	Type of Rope			
	$\frac{5}{8}$" Hemp	$\frac{1}{2}$" Dacron	$\frac{5}{8}$" Polypro[1]	$\frac{5}{8}$" AB/PNX[2]
One Wrap	119	95	110	120
Two Wraps	221	184	215	257
Munter Hitch	300[4]	182	300[5]	240
Carabiner Brake	300	119	123	270
Carabiner One-Wrap[3]	250	120	148	199
Figure Eight	300[4]	238	294	300[5]

[1] Once slippage began, it continued at a load one-half to three-quarters of the load listed.

[2] American Brand PNX Rope is a three-strand rope with each strand made of a core of monofilament polypropylene covered by alternate yarns of multifilament polyester and monofilament polyester.

[3] This piece is designed for ropes in the 8mm to 11mm range and is too small for the ropes used.

[4] These supported loads well over 300 pounds. (Grasping the "load" end of the system, the tester lifted himself off the floor without overcoming the brake.)

[5] These supported loads until slightly over 300 pounds. (Slippage began as the tester lifted himself off the floor.)

Through the years, many configurations of performance and audience areas have evolved within theatres in order to provide a better relationship between the audience and performer. Today, in many theatres, these configurations are meant to be flexible and are reestablished for each production. Each individual arrangement should be reviewed with the provisions of fire safety codes in mind. This article provides a checklist to maintain the public's safety during theatrical performances, particularly whenever the seating layout is temporary. For the most part, these are common-sense issues.

SEATING

Though the local fire authority usually posts a "maximum occupant load" for each theatre space, its true capacity is dependent on several factors, and the following points should be scrupulously observed:

AISLE SEATING	The back-to-back spacing of rows must be at least 32", provided the back-to-front spacing is at least 12".
CONTINENTAL SEATING	If there are more than 14 seats between aisles or 7 seats between an aisle and a wall or other such barrier, then an additional $\frac{1}{4}$" must be added to the row depth for each seat added to the row.
PEW (BLEACHER) SEATING	When seats have no backs, the depth between rows must be at least 22". When there are no dividing arms between chairs, at least 18" must be provided along the bench for each person.
CABARET SEATING	When fixed and/or loose chairs and tables are used, a clear path no more than 10' long must be provided from each seat to an aisle.

ACCESS FOR THE DISABLED

In every place of public assembly, access must be provided for the disabled. Therefore, spaces need to be provided for wheelchairs. Two wheelchairs must be accommodated for the first 100 seats, with one additional space for every 100 seats thereafter.

AISLES

Since an aisle's primary purpose is to get the audience to and from their seats, the aisles must be kept clear of scenic and other obstructions at all times. Additionally, nothing should hang lower than 8'-0" over any aisle used by the public. When an aisle leads to an exit at both ends, the width of the aisle should remain the same along its length. In no case should an aisle's width decrease as it approaches an exit. Additionally, all aisles longer than 20' must end at a cross-aisle or door leading to an exit from the building. If there is seating on both sides of an aisle, the aisle width must be at least 42". Aisles that have seating on one side only must be at least 36" wide.

EXITS

Two exits are required from performance spaces seating 50 to 300 people; three, for those seating 300 to 1000 people; and four, for those seating more than 1000 people. These exits should be as remote from each other as possible and must lead to independent exits from the building. The doors in these exits must open out from the space. Whenever the space is occupied by the public, the doors must be unlocked and kept clear of all obstructions.

Every fire exit shall be so marked with an illuminated exit sign. Each audience member must be able to see at least one of these signs at all times. Since the signs are intended to direct the audience from the building, the signs must be placed perpendicular to the path of travel and at a reasonable height. Every exit sign is to remain lit at all times when the space is occupied.

ILLUMINATION

As is true of exit signs, all aisles and audience stairways must be illuminated while the audience is in the space. In the event of an emergency or power failure, exit signs and aisle lights must remain lit for at least one and one-half hours.

FLAMEPROOFING

While every state has different regulations concerning the flameproofing of scenery, it is wise to flameproof all the scenery whenever there is no wall of protection between the audience and a set. These precautions apply not only to the visible, onstage, sides of the walls and floor, but to the off-stage and backstage portions of a set as well, and particularly to familiar objects and materials that are used atypically. Some carpets, for instance, may be flame-resistant while in place on a floor, but hanging them as tapestries greatly decreases their fire resistance.

FLAME ONSTAGE AND PYROTECHNICS

Whenever there is flame onstage — even cigarettes or candles — a fire extinguisher should be kept nearby. To provide the maximum safety for the audience, any flame used onstage should be kept as far away from the audience as possible. All flammable materials to be used near the flame should also be flameproofed. Pyrotechnic displays, which are not the same as flame, are often stringently regulated. Check with your local fire authority for specific regulations in your area.

AUDIENCE AWARENESS

Before any performance, the audience should be informed of the intention to use any special effects that are known to affect some individuals adversely. Generally, clear notice that a production involves the use of fog, smoke, pyrotechnics, strobe lights, and/or other potential nuisances or hazards should be posted in the lobby and should appear in the program.

FIRE SPEECH

Whenever a space's initially intended configuration is altered, a pre-show announcement should be made. The "Fire Speech" should contain a reminder of the prohibition of smoking, should clearly define the routes of exit, and should reiterate the warnings concerning special effects and/or chemicals to be used during the performance.

FINAL NOTE

All of the above guidelines are based on the National Fire Protection Association's *Life Safety Code* (NFPA-101). Whether the code is more lenient or stricter than your city and state laws, these guide-lines represent a starting point. It is up to designers and, ultimately, to technicians and technical managers to insure the safety of the audience at all times. When you are presented with a new seating layout for a production or know that a production will use special effects, call your local fire authority to verify the regulations that pertain to your theatre.

Communication with the fire authority should always be maintained. No matter how simple you believe your production to be, it is wise to discuss the plans for your production as soon as you have an idea about the design. The more communication between you and the fire authority — and the earlier it occurs — the smoother and more pleasant the production process will be for everyone involved.

Additional information may be obtained from the NFPA's Code, Chapters 5, 8, and 9. You may con-tact the NFPA at

> National Fire Protection Association
> Batterymarch Park
> Quincy, MA 02169

EDITORS' NOTE

Laws governing fire safety have changed since this article's original publication, and vary from one locality to the next. Consult with state and local fire authorities for information about those regu-lations currently in effect in your area.

❧❧❧

TECHNICAL BRIEF

Scenery

MATERIALS

Burlap-Backed Carpet (we salvaged ours from an old apartment building)
Swift's 3917® Flexible Glue
Dental Plaster
Sawdust
Tinting Agent
3mil Polystyrene

EQUIPMENT

Cutawl®
Buckets

Cover the area to be bricked with the carpet, burlap side up. Use a Cutawl® to cut a convenient size stencil from the polystyrene. We worked with a stencil 5'-0" x 3'-4". Then mix the "cement" by adding equal parts of thinned flexible glue (9 parts glue to 1 part water) and dental plaster. If tinting is desired, add the tinting agent to the thinned glue first. Next add sawdust to the mixture at approximately a 4:1 ratio, depending on the density of sawdust. The cement is then poured into the stencils and scraped down with 1x4 sticks.

Two problems occurred. First, the mix dries quickly, and cleaning the stencils proved difficult. We had to coordinate our personnel so that the mixing rate and the pouring/scraping process could be continuous. Second, a pronounced color shift occurs when the plaster is added, so experimentation is needed to achieve the desired color. Color control was finally achieved by mixing larger batches of the tinted thinned glue.

The results exceeded our expectations. The bricks were $\frac{3}{16}$" deep with well-defined grout lines. To determine the quantity of mix you will need, estimate one gallon of finished mix to cover approximately 7 square feet. This figure will vary with the thickness of bricks desired.

꿏꿏꿏꿏

A decision tree is a conceptual model used to make decisions under conditions of uncertainty. A decision involves a choice between several strategies, each of which may result in several possible outcomes. An outcome is any possible situation that may occur, either by choice or because of chance events. When confronted with a decision for which there is insufficient information, one intuitively weighs the possible outcomes according to their likelihood and chooses the one that seems best. The decision tree is a method of formalizing this intuitive decision-making process and keeping track of complicated situations.

PROBABILITY

Uncertain situations may arise either from random events or from lack of information. The key to dealing with uncertain situations is the evaluation of each possible outcome in terms of its probability. A probability is a number between 0 and 1 indicating the likelihood that a particular event will occur. A probability of 0 indicates a certainty that an event will not occur; a probability of 1 indicates a certainty that an event will occur. The probability of a coin landing tails up is 0.5; the probability of a six-sided die landing with any given face showing is one-sixth or 0.166. The probability that an event will not occur is equal to 1 minus the probability that the event will occur.

A probability for a given event may be objective or subjective. An objective probability is determined statistically. A subjective probability is derived from a subjective assessment of the likelihood that an event will occur. Objective probabilities are preferred but rarely available, so decisions are usually made on the basis of subjective probabilities.

Decision analysis requires that events be mutually exclusive and collectively exhaustive. An event cannot have two different outcomes simultaneously, and an event must include all possible outcomes. The latter means that the probabilities of all the outcomes must add to a total of 1. Two important rules for probabilities are as follows:

1.   The OR rule: $p(A \text{ or } B) = p(A) + p(B)$. The probability of any of several mutually exclusive events occurring is equal to the sum of the probabilities of the individual events.

2.   The AND rule: $p(A \text{ and } B) = p(A) \times p(B)$. The probability of A and B both occurring is equal to the probability of A times the probability of B. A and B must be independent events, one following the other.

EXPECTED VALUE

The decision tree uses probabilities to calculate Expected Cost (EC) or Expected Value (EV). The EC of an event is the real cost times the probability that the event will occur. The EV is the real value times the probability that it will occur. For example, if I offer to flip a coin and give you $1.00 if it comes up heads your EV is $p(\text{heads}) \times \$1.00 + p(\text{tails}) \times \$0$, or $(.5 \times \$1.00) + (.5 \times \$0.00) = \$0.50$. Note that the case where the coin comes up tails is accounted for. All the possible outcomes must be considered: the probabilities must add to a total of one.

TREES

There are often many alternatives in a decision that depend upon other decisions and chance events. These are presented as a tree where each branch represents a process or action. The places where the branches split are called "nodes." Nodes indicate either decisions (usually drawn as square boxes) or chance events (usually drawn as circles). The outcomes are placed at the ends of the branches. Simple trees can be placed end to end so that an outcome of one tree provides the starting node of the next. The decision tree proceeds like a timeline from left to right. Once a decision has been made or a chance event has occurred, there is no going back.

For example, let us suppose that you are in a position to buy a welder. To simplify the example we will use a one-year planning horizon. The options are to buy a new welder for $300.00, to buy a used machine for $200.00, or to rent a machine for $60.00 a week when needed. The new machine carries a warranty, so there will be no repair costs. The used machine, according to the local welding shop, will cost about $150.00 to repair if it breaks, and they give 50/50 odds on its breaking. In addition, if the used machine breaks you may rent a machine instead of fixing it. You estimate that there is a 0.6 probability that it will be used for 4 weeks, a 0.3 probability that it will be used for 8 weeks, and a 0.1 probability that it will not be needed at all. Note that the probabilities are mutually exclusive and collectively exhaustive. The options are graphically presented in Figure 1.

EVALUATING THE DECISION

Once the structure of the tree has been drawn, the relevant information can be added. First, the costs of various actions or events are added to the appropriate branches. Second, the outcomes are determined by starting at each endpoint of the tree, summing costs along the way. The resulting cost (or value) of each outcome is placed at the endpoint corresponding to that outcome. Third, the probabilities for the uncertain events are added to the branches growing from the chance nodes.

The result of this process for the example is shown in Figure 2. The costs are written along the branches where they occur and the probabilities are listed in parentheses for the branches following each chance node. The EV or EC is determined through a process known as "averaging out" or "folding back." Each node of the tree can be looked at as the root of a simple tree, so starting at the right-hand side of the tree we look at the first simple tree.

If the first node is a decision node, we can choose the best outcome: if we are minimizing costs, we choose the lowest cost; if we are maximizing value, we choose the highest value. We then place the value chosen next to the decision node where it will become the outcome for the next part of the tree. In the example, the rightmost nodes in the "buy used" branch are decision nodes. In Figure 2 the more costly outcome of these nodes has been crossed off, and the less costly outcome, $350.00, transferred to the node to be used for the next calculation.

If the node is a chance node, we evaluate it by calculating its EV or EC. We calculate the sum of each of the outcomes for that node multiplied by their respective probabilities. The resulting value is written next to the chance node, and it again becomes the outcome for the next set of calculations. In the example, the rightmost nodes for both the "new" option and the "rent" option are chance nodes. In the "new" case, all outcomes are identical, and since the probabilities for each chance node must total 1, the EC is the same as the outcomes: (300 x .3) + (300 x .6) + (300 x .1) = $300. In the "rent" case, the EC is (480 x .3) + (240 x .6) + (0 x .1) = $288.00.

This process is repeated until we come to the left-hand side of the tree. In the example, the outcomes from the decision nodes at the right-hand side of the "buy used" branch and the $200.00 "doesn't break" outcome are combined to form the EC at the three chance nodes. These, in turn, are used to calculate an EC for the "buy used" branch as a whole. This EC is (275 x .3) + (275 x .6) + (200 x .1) = $267.50. The three main branches now have ECs that can be compared. The "buy used" option has the lowest EC and so is the most economical option.

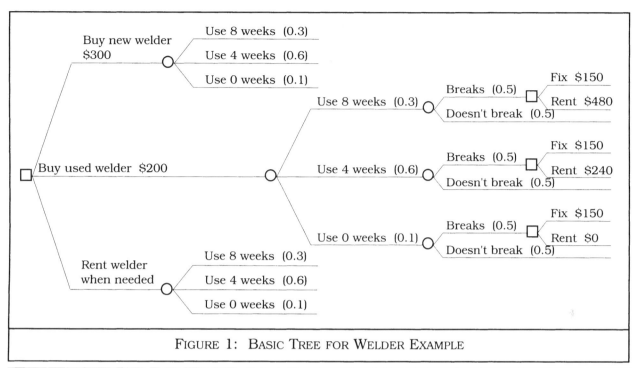

FIGURE 1: BASIC TREE FOR WELDER EXAMPLE

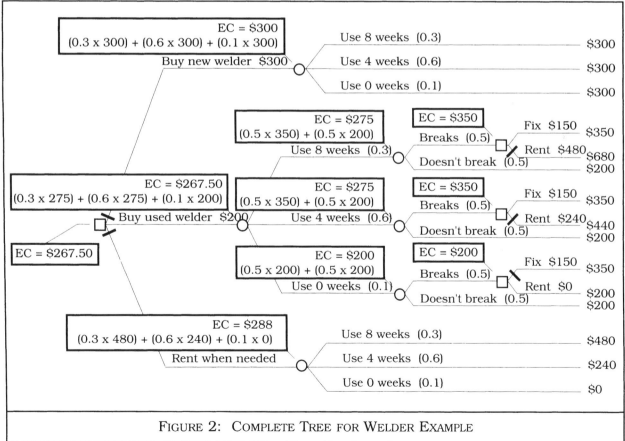

FIGURE 2: COMPLETE TREE FOR WELDER EXAMPLE

Using back-flap hinges to join two standard-frame flats into stock masking units results in three problems. First, the crack at the hinge line between the flats may or may not remain masked, depending on how much the hinged styles warp over time. Second, the projecting cornerplates and keystones used on standard flats are a nuisance. Making each folded unit as much as one-third thicker than it needs to be, they eat up storage space at a surprising rate. And, though they offer convenient hand-holds, they also pose a tearing threat to the fabric surfaces of stock flats, which undergo repeated handling. Finally, because the units typically fold one way only, they are seldom built to be more than two flats wide.

The stock masking units described in this article use a cloth version of the paper hinge found on many Oriental screens. The use of this hinge provides complete hinge-line masking and allows units to be made more than two flats wide while minimizing their thickness to facilitate storage and shipping.

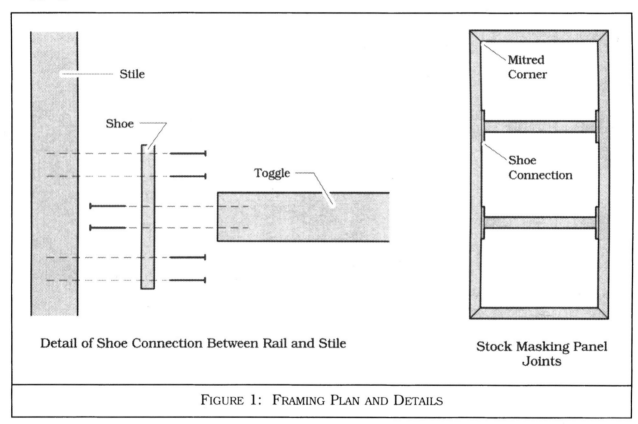

Detail of Shoe Connection Between Rail and Stile   Stock Masking Panel
                                                                         Joints

FIGURE 1: FRAMING PLAN AND DETAILS

FRAMING AND COVERING

The stiles of the individual flats are connected to the rails by mitre joints. Shoes measuring $\frac{3}{4}$" x 8" are nailed to the ends of the toggles, and the "shoed" toggles are then nailed between the stiles. See Figure 1. Coated, withdrawal-resistant nails are used in all joints. Though the use of half-laps at the rails and toggles may seem feasible, the author finds the joints described above to be superior because, unlike half-laps, they do not cut into the stiles. After the frames' sharp edges have been sanded or planed off, the surface fabric is laid out on, wrapped around the edges of, and stapled to the frame at the rear. Though muslin, duvetyne, and no-wale corduroy are all suitable as cover fabrics for these flats, the author prefers the denser body of Veltex.

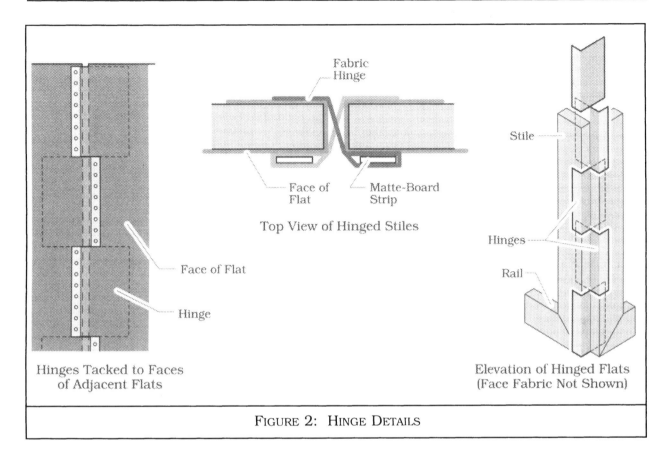

FIGURE 2: HINGE DETAILS

HINGING

Frames to be hinged are laid face up next to each other. Next, 4" x 6" strips of the same fabric used for the surface are laid face down on the stiles to be hinged as shown in Figure 2 and tacked down through $\frac{5}{8}$" x $5\frac{7}{8}$" strips of matte board or other heavy cardboard. The frames are then turned face down with hinge stiles abutting each other, and each strip of fabric is pulled up through the hinge joint. After the gap at the hinge joint has been minimized, the fabric strips are pulled snug against the stile of the "partner" flat and tacked in place to form a durable continuous hinge.

NOTES

Since these hinges permit frames to fold either face to face or back to back, masking units may be made more than two flats wide. The author has found three-flat units to be the type most useful in mounting productions in "found" spaces.

The addition of a strip of plastic to the top and bottom edges of each finished unit, though not essential, will prevent fabric wear. The reader is cautioned that these frames cannot be slid on their hinged edges without damage to the hinges.

❧❧❧

Mass Producing Styrofoam® Balusters

Identical balusters can be mass produced economically for stage use by turning high-density Styrofoam® on a drill press. While turning wood requires expensive material and skilled labor, this method replaces the wood with inexpensive Styrofoam® and simplifies turning by using a profile pattern. The task is accomplished in three steps: building the Styrofoam® blanks, building the profile pattern, and turning the blank on a drill press.

BUILDING THE STYROFOAM® BLANK

1. Cut a $\frac{3}{8}$" dowel three inches longer than the baluster and chamfer one end. The dowel later serves as a centering device during installation. To help secure the foam to the dowel, imbed appropriately sized finish nails into the dowel from four directions at the points at which the baluster is to be the widest. To accommodate the dowel, the center two sections of foam must have $\frac{3}{8}$" semicircular grooves routed down their lengths.

2. Laminate commercial-grade polystyrene foamboard ("blue foam") with an aqueous neoprene latex adhesive such as 3M's Fastbond 30®. Avoid mastic adhesives, which do not dry completely, resulting in delamination during turning. Laminate the foam to the dowel 2" from the top. Trimming the laminated block to a hexagonal shape decreases turning time.

3. The size of the finished baluster is limited by the drill press dimensions. The overarm (column to chuck center) dimension limits the baluster thickness. Baluster length is limited by the extent of table travel and/or excessive chatter that will occur when turning too slender cross sections.

BUILDING THE PROFILE PATTERN

1. The pattern is a positive profile of the finished baluster. Cut two profile patterns simultaneously from $\frac{1}{2}$" plywood. This ensures that the profiles are identical and that the balusters will be symmetrical. Allowing the profile to extend an inch or two beyond the length of the baluster will make it easier to shape the baluster ends.

2. Attach the profile sides baseplate of $\frac{3}{4}$" plywood and a back of $\frac{1}{2}$" plywood. The assembled pattern should be bolted to the drill press table with carriage bolts and to the column with a U-bolt.

3. Install a bushing at the centerline of the profile in the base plate to provide a pivot point to minimize spindle chatter during turning. See Figure 1C. The heat generated during turning precludes the use of low-temperature plastics or wood for the bushing. Therefore, the bushing must be made of a long-lasting material, such as Teflon® or nylon.

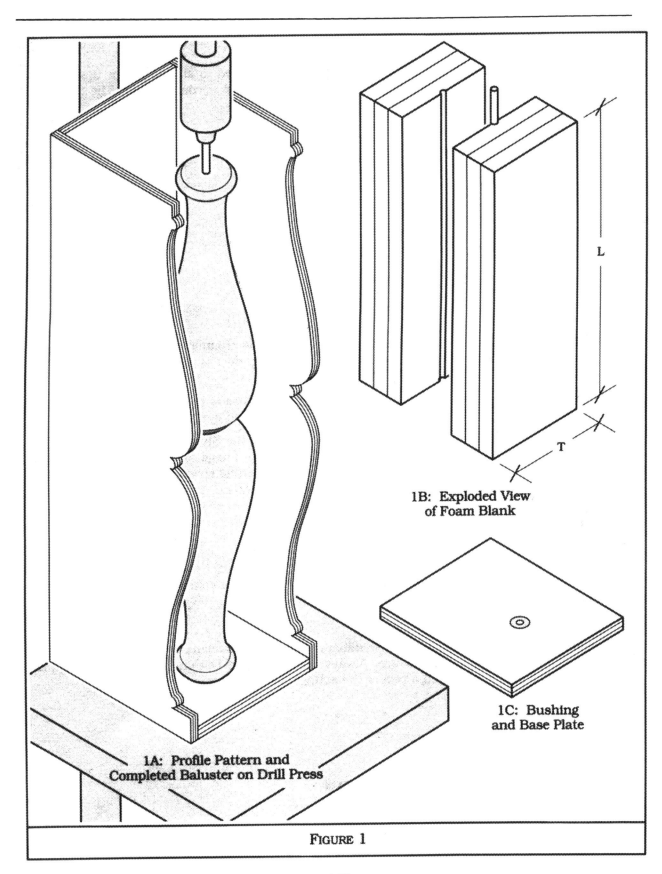

1B: Exploded View
of Foam Blank

1C: Bushing
and Base Plate

1A: Profile Pattern and
Completed Baluster on Drill Press

FIGURE 1

TURNING THE BLANK

Suitable carving tools are easy to make, but they must be designed to abrade rather than cut or gouge the Styrofoam®. Gouging foam at drill press speeds would be dangerous. The following would comprise a renewable set of carving tools:

FOR ROUGH FORMING

$1\frac{1}{8}$" closet rod with 60 grit sandpaper

FOR IMAGING

$\frac{7}{16}$" dowel with 60 grit sandpaper
$\frac{7}{16}$" dowel with 100 grit sandpaper
$1\frac{1}{4}$" x $\frac{1}{8}$" stick with 150 grit sandpaper

FOR FINE DETAIL LINES

$1\frac{1}{4}$" x $\frac{1}{8}$" stick beveled to zero

1. Load the 2" end of the dowel into the drill chuck. Seat the chamfered end into the bushing. To minimize friction between the dowel and the bushing, do not apply downward pressure while tightening the chuck jaws.

2. Operate the drill press between 1750 and 2750 rpm. Because of the generation of fumes and dust, work in a well-ventilated area and wear a respirator and eye protection.

3. Hold the carving tool horizontally with both hands. Form the Styrofoam® blank into a cylinder by using long vertical strokes. Next, follow the pattern image to produce a rough shape. Greater detail is achieved by using smaller tools for succeeding strokes. Remove the finished baluster. Clear the debris. Set up the next blank.

FINAL NOTES

A preliminary limited test run will help determine specific requirements. In production runs, balusters can be completed in less than ten minutes each. Setting up a comfortable work station and preparing the blanks properly will increase the system's efficiency. In general, this system produces inexpensive balusters quickly and safely.

Users should note, however, that cutting Styrofoam® releases hazardous gases and can generate enough heat that there is a potential for fire. Always wear protective clothing, gloves, and an appropriately filtered respirator, and keep a proper fire extinguisher ready nearby.

SOURCE

Clayton Austin.

In planning the layout for any deck, the technical director seeks to minimize costs by using as many stock platforms as possible and to minimize build and load-in time by devising the simplest possible legging system. If the deck is to be level, the technical director can simply draw the layout on an overlay of the set designer's plan. But laying out a raked deck is a more difficult problem since the designer's plan does not represent the true size and shape of the deck.

This article outlines a solution to laying out a deck that is raked and also oddly shaped. The method, which requires a copy of the designer's deck plan and a clean piece of vellum, has three major components: determining a rake axis line, dividing the rake axis line into useful segments, and orienting standard platforms along the rake axis line. The following example assumes the use of stock platforms measuring 4'-0" x 8'-0" and 4'-0" x 4'-0".

DETERMINING A RAKE AXIS LINE

To define the rake axis, identify the deck's three most critical points. The points do not necessarily include the highest and lowest points on the deck. For example, they may instead be points at which the deck comes into contact with an escape. Once they have been marked, the rake axis can be determined with the following method.

1. Draw baseline **AB** connecting the highest and lowest critical points (points **A** and **B**, respectively, in Figure 1A).

2. On baseline **AB**, construct a section through the deck, section **ABB'A'**. See Figure 1B.

3. Locate point **C'**, the middle critical point, at its correct elevation on the section, and construct line **C'D** perpendicular to baseline **AB**. See Figure 1B.

4. Draw line **CD**. See Figure 1C. Note that all points on line **CD** are the same height off the floor.

5. Finally, erect rake axis line **XX'**, perpendicular to line **CD** at any convenient point along **CD**. See Figure 1C.

DIVIDING THE RAKE AXIS LINE INTO USEFUL SEGMENTS

Because this deck is raked rather than level, its plan view does not show its true size and shape. Any line drawn perpendicular to the rake axis line can be scaled into segments, since all the points it includes are at a single, constant elevation. Lines drawn in any other orientation, however, cannot be scaled, since they represent changing elevations. Thus, in order to divide the rake axis line into meaningful segments, the technical director must use the following approach.

1. Erect perpendiculars from points **A** and **B** through the rake axis line. See Figure 1D.

2. Construct a new section through the deck, section **MNN'M'**, using a segment (**MN**) of the rake axis line as a new baseline. See Figure 1D.

3. Starting at **M'**, scale line **M'N'** at 4'-0" intervals and erect lines perpendicular to the rake axis line that extend in both directions as in Figure 1E.

4. Lay the clean vellum over the plan and trace both the rake axis line and the perpendiculars drawn in the preceding step.

5. Starting at the rake axis line, scale the outermost perpendiculars at 4'-0" intervals, and connect the interval marks to produce a grid of rectangles as in Figure 1F. (Note: the gentler the rake, the more closely these rectangles will resemble squares.)

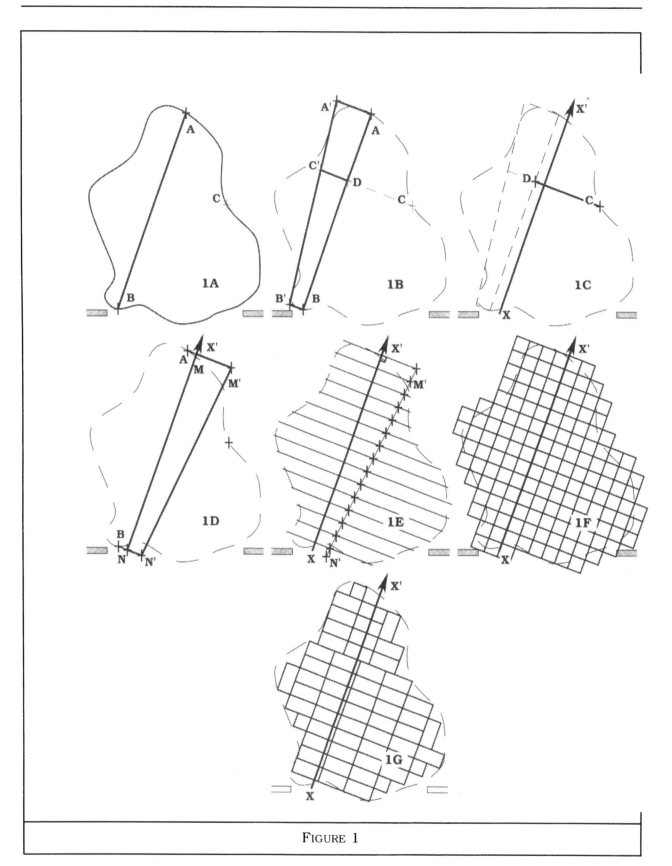

FIGURE 1

ORIENTING STANDARD PLATFORMS ALONG THE RAKE AXIS LINE

1. Keeping the rake axis line drawn on the vellum parallel to that drawn on the plan, slide the vellum across the plan until the greatest possible number of complete rectangles fits within the outline of the deck.

2. Tape the vellum in place, trace the outline of the deck, and indicate the best placement of the stock platforms. Such a layout appears in Figure 1G.

NOTES ON LEGGING

This raked deck layout approach saves time in devising a legging system. Because each line of cross-axis platform edges represents a constant elevation, it can be supported by a series of legs cut to equal lengths or by a single stud-wall frame.

❧❧❧

For a production of *A Streetcar Named Desire* at James Madison University, Set Designer R. Lee Kennedy wanted a freestanding staircase that would have the look of New Orleans ironwork. It was also to be an open-carriage design, *i.e.*, it could have no external support other than the stage floor and a platform at the upper end. In addition, actors needed clear passage under it, and Lighting Designer Mark Darden wanted a clear view of the unit silhouetted against the cyc.

JMU students Kim Sprouse and Joel Moritz, who had taken on the project as part of an Advanced Technical Theatre course, found little help from theatrical source books: in them, Sprouse and Moritz found only closed-carriage designs, which hide a series of trestle tread supports (gates) behind flats. Sweet's Catalog was somewhat more helpful because it contains many architectural units that met the designers' criteria. Predictably, though, all of the commercially available units were far too costly for the production budget. By drawing inferences from the Sweet's Catalog descriptions and through discussions with commercial staircase manufacturers, however, they gleaned enough information to design the shop-built, open, curved staircase shown in Figure 1.

The staircase was built around a 7'-0" inside radius. Its treads were 3'-0" wide and measured 9" along the inside stringer and 1'-1" along the outside stringer. It was 6'-8" tall and comprised ten treads built up on an 8" rise. It was designed to carry the weight of two average-size actors who, it was assumed, might happen to climb the staircase simultaneously — a total combined static and dynamic load somewhat in excess of 400 pounds and far less than the 1000 pounds on which some manufacturers claim to base their commercial designs. Sprouse and Moritz judged that this load could probably be carried by a staircase that used only two carriages made of $\frac{1}{4}$" x 12" steel bar, even if they were cut to reveal the outline of the treads and risers. The design also used $\frac{3}{4}$" plywood treads supported by $\frac{1}{4}$" x 2" x 2" angle iron welded between the carriages at each tread/riser intersection as shown.

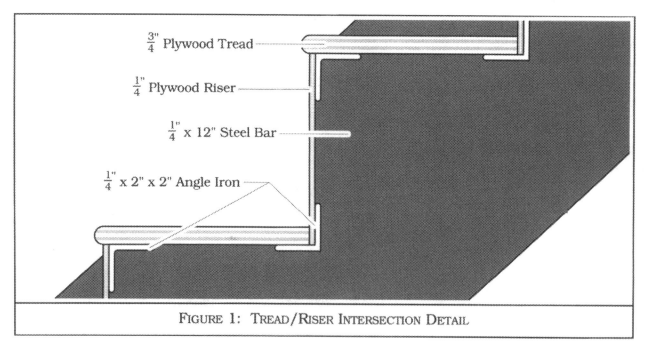

$\frac{3}{4}$" Plywood Tread

$\frac{1}{4}$" Plywood Riser

$\frac{1}{4}$" x 12" Steel Bar

$\frac{1}{4}$" x 2" x 2" Angle Iron

FIGURE 1: TREAD/RISER INTERSECTION DETAIL

CONSTRUCTION AND ASSEMBLY

Since the JMU Theatre shop has no bending tools and only limited metal-working equipment, the construction method was restricted to the possibilities offered by a standard power hacksaw and buzz-box AC welder. First, a half-scale cardboard mockup was produced to test the dimensions on the designer's drawings. This process also allowed us to test the approach of cutting the carriages from flat material and then bending them. Satisfied with the model, we began construction.

After the carriages had been cut to length and notched to receive the treads and risers, we welded the topmost and bottommost pieces of angle iron in place. Next, we leaned the stringers against the back wall of the theatre and attached the topmost angle iron to the wall at the appropriate height. Then, after swinging the bottom angle iron into its proper orientation and securing it to the stage floor, we bent the carriages into the final desired curve by hand, inserting tread-length 2x4s between them at each tread location in order to establish and maintain correct spacing between the stringers. This done, we replaced each 2x4 by welding an angle iron strut in place, working from top to bottom. When we disconnected the unit from the wall and floor, the stringers remained curved, and we had only to attach the treads, risers, and handrails.

We bolted the pre-cut treads in place, and followed them with a series of $\frac{1}{4}$"-plywood risers that had been cut out to resemble New Orleans gingerbread. The railing consisted of 1" square tube steel balusters (welded to the upstage/stage left side of each carriage) and handrails of strap steel.

❧❧❧❧

I-Beam Tracking System

Edmund B. Fisher &
Thomas P. Sullivan

A production at the Juilliard School required that a heavy wagon be pushed across a steeply raked deck, from offstage left to centerstage, where it was rolled onto an elevator. The full-stage show deck was raked at about 2" per foot of run. The top deck of the 8' x 10' x 8' tall wagon had to fit flush with the show deck when the elevator was lowered. This meant that tracking and placement of the wagon on the elevator was critical.

THE PROBLEM

Platforms that are rolled across a raked deck tend to drift down the rake as they travel. Even though a knife guide will limit the amount of drift, the knife will inevitably rub along the guide slot. The friction that our 1700# load would cause was considered a significant problem, for in addition to requiring extra force to move the wagon, it would create excessive noise and the possibility of jamming was significant.

THE SOLUTION

We realized that our biggest problems were the result of one factor — the raked deck — and that if we could get around the rake these problems would be eliminated. The solution was simple enough: using a level track for the wagon underneath the show deck would eliminate the rake-induced tendency to drift. We soon decided that an I-beam section would be an ideal wagon-caster undercarriage, with its web serving effectively as an unobtrusive knife guide and its flanges as mounting surfaces for both wagon and casters. See Figure 1.

TRACK

Our track was built as a narrow stressed-skin panel with a $\frac{3}{4}$"-plywood top, $\frac{5}{4}$ x 4 framing, and a $\frac{1}{4}$" plywood bottom. It was built in 16' sections with male and female ends to assure rigid connection and tight-fitting seams between sections. The track was built narrower than the space in the studwall in which it rested and yet wider than the wheels that rolled on it, so that alignment of the track with the slot in the deck was not critical. The track was screwed to the studwalls to prevent shifting. Unpainted, tempered masonite was added on top with the seams offset from the track seams to provide the smoothest possible surface for the casters to roll on.

UNDERCARRIAGE

An I-beam has the perfect cross section for this application. We opted to use two I-beams that ran the full length of our wagon because they would present the smallest number of leading edges to catch on seams in the deck and they could be installed quickly. Further, full-length I-beams would also be able to bear the weight of the wagon on a minimum number of casters without deflecting appreciably. Caster mounting and alignment is important, of course, to ensure that the carriage travels in a straight line. We found that, with the casters properly aligned, the I-beam web never rubbed against the edges of the slots.

DECK

Thin enough to provide clearance for the I-beam, the deck at Juilliard was only $1\frac{1}{2}$" thick. After the deck was installed, two parallel $\frac{1}{2}$" slots were cut, centered above the track, for the I-beams to run in. Since the tracks were 6" wide, there was about $2\frac{3}{4}$" of unsupported deck on either side of the slots, which was important to take into account when laying out the plywood layer of the deck.

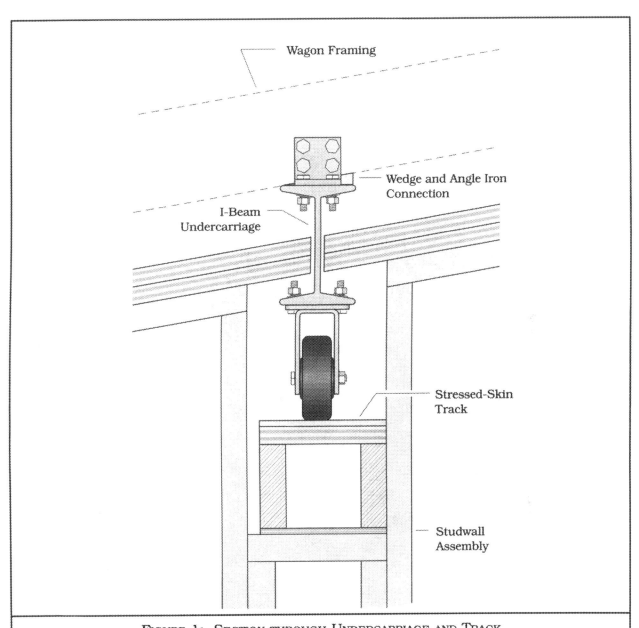

FIGURE 1: SECTION THROUGH UNDERCARRIAGE AND TRACK

CONCLUSION

This system provides a tracking wagon that is easy to move, very quiet (since the moving parts are below the deck), and very accurate. We suffered no noticeable drift either upstage or downstage during the run, and the two $\frac{1}{2}$" slots in the deck were almost unnoticeable.

❧❧❧

A Measured Approach to Kerfing

Daniel J. Culhane

The typical trial-and-error approach to kerfing materials that are to be bent into arcs often produces less than desirable results, whether the objective is to preserve the maximum amount of material or to form an arc as quickly as possible. The following method, which I learned from master craftsmen at the Children's Theatre Company, can help the builder reach both objectives more directly. First, determine the outside radius of the required arc. Measure and mark the length of that out-

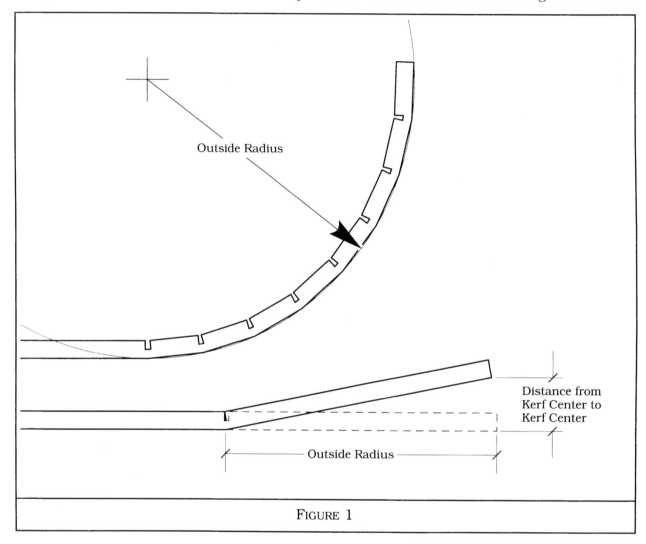

Outside Radius

Distance from Kerf Center to Kerf Center

Outside Radius

FIGURE 1

side radius along one end of a scrap piece of the stock to be used. Kerf the stock at the center of this mark. Place the stock on a flat surface and raise the measured end until the saw kerf closes, as in Figure 1.

Measure the vertical distance from the flat surface to the bottom of the stock. This measurement is the distance needed from kerf center to kerf center in order to approximate quite closely the required radius.

This approach allows a smooth and uniform arc to be formed in any material and cross section that will accept a kerf.

In performance spaces that seat the audience close to the stage, as with thrust or arena settings, spike marks of glow tape, colored plastic tape, or paint are distracting on a stage floor; and tape or paint marks virtually disappear under the low-light/no-light scene change conditions which are often used in such spaces.

ALTERNATIVE 1: LEDs

One alternative to the usual spike marks is the use of LEDs set into the floor. (See the *Technical Brief* article, "Two Easy LED Circuits" by Tom Neville.) In a Yale Repertory Theatre production of *Troilus and Cressida*, narrow plywood bridges across a trap-room-deep moat connected the wings to the central acting area. Holes were drilled down the centerlines of the bridges, and LEDs were inserted from below, creating a runway effect while remaining unseen by the audience.

LEDs of different colors can signal the status of traps in a darkened stage. In the Yale Dramatic Association's production of *Sweeney Todd*, a pneumatic lift was used to lower a trap into the stage floor. The edge of the trap was marked with two circuits of LEDs, one green and one red. Before the lift could be lowered, the operator beneath the stage had to unlock the trap platform and throw a switch that turned off the green LEDs and illuminated the red LEDs. Thus, actors moving toward the trap in a dim or smoke-filled scene found the trap easily and knew the status of the lift platform.

ALTERNATIVE 2: AMBIENT LIGHT FROM BELOW

A second alternative to tape was discovered accidentally in the process of mounting a Yale School of Drama production of *Romeo and Juliet*, in which a rolling bed had to be brought onstage in a blackout. The bed was secured to a shiny black deck with steel pins that slid into holes drilled in the floor. As carpenters drilled holes in the deck for the pins, it became obvious that light from the trap room below would create bright white spikes that would not be visible from the audience ten feet away. Once the ambient light in the trap room had been reinforced by the addition of a scoop, the bed was effectively spiked.

❦❦❦❦

TECHNICAL BRIEF

Scenery Decks

Stressed-skin platforms are a variation on stressed-skin panels used in building construction. They are especially useful for long spans where heavy beams and numerous supporting legs are visually or logistically undesirable.

A stressed-skin platform consists of stringers that run the full length of the span and are covered on both sides with plywood. Lumber headers are placed across the stringers at each end of the panel. If the platform is longer than the plywood being used, the plywood butt joints must be backed with plywood splice plates centered on the joint and backed with lumber blocking. The plates that fit between the stringers should be as thick as the skin being spliced. The length of the splice plate varies with the thickness of the plywood: 6" long for $\frac{1}{4}$" plywood; 8" long for $\frac{5}{16}$" plywood; 10" long for $\frac{3}{8}$" plywood; 12" long for $\frac{1}{2}$" plywood; 16" long for $\frac{5}{8}$" or $\frac{3}{4}$" plywood. All joints must be glued and nailed.

The design principle employed is similar to that of an I-beam. The stringers, like the web of an I-beam, provide resistance to horizontal shear. The top and bottom skins, like flanges, provide resistance to compressive and tensile stresses caused by bending. The top skin must also provide adequate support between the stringers and is usually $\frac{5}{8}$" plywood or thicker.

The spans listed in the table below assume 4'-wide platforms, continuous stringers 16" on center, $\frac{5}{8}$" plywood as the top skin, and $\frac{3}{8}$" plywood as the bottom skin.

STRINGERS AND MAXIMUM SPANS Based on a 50psf Loading Condition	
Nominal Dimension	Maximum Span (Failure in Bending)
1x4	8'-0"
1x6	12'-0"
2x3	8'-0"
2x4	10'-0"
2x6	14'-0"
2x8	17'-0"
2x10	20'-0"
2x12	23'-0"

NOTES

For more exact specifications and formulas for designing stressed-skin platforms, consult the following

Plywood Design Specifications, Supplement #3
American Plywood Association
1119A Street
Tacoma, WA 98401

This article describes an easy method to elevate, support, and secure platforms without the use of hardware. The method employs wood modules — boxes with protruding corner cleats — that can be quickly fabricated providing the user with height and layout flexibility. See Figure 1.

The symmetrical cross section allows modules to be interchanged with one another. The snug fit of the platform framing between the cleats holds the platforms together without the use of joining hardware. Four adjoining platform corners can be legged and held together by placing each corner over a single cleat. See Figure 2. Two platform edges can be legged and held together by placing each over two cleats. One platform corner is legged by covering all four cleats. In all cases, the module is not clamped to the platform.

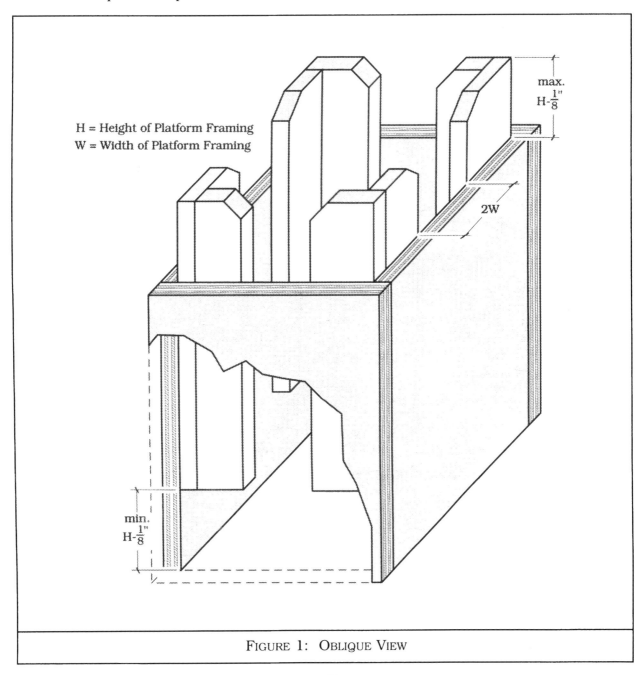

max.
$H-\frac{1}{8}"$

H = Height of Platform Framing
W = Width of Platform Framing

2W

min.
$H-\frac{1}{8}"$

FIGURE 1: OBLIQUE VIEW

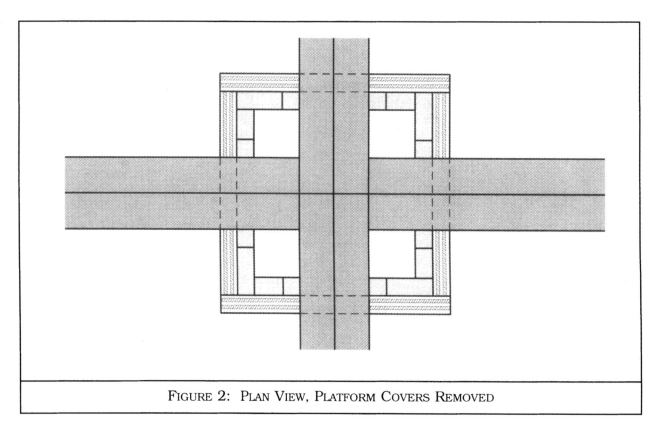

FIGURE 2: PLAN VIEW, PLATFORM COVERS REMOVED

Platform framing dimensions determine the modules' size. The distance between cleats is twice the platform framing width (2W). Box height is determined by subtracting the platform thickness from the desired height. The length of the cleat above the box should allow the platform framing to rest on the box (H minus $\frac{1}{8}$"). Cleats are beveled for easy assembly.

Holding cleats back from the bottom of the box the same distance they protrude from the top permits them to be stacked for use and storage.

OTHER NOTES

1.   Design height increments to suit needs.

2.   Always lift platforms straight up to avoid framing damage.

3.   Non-skid pads on the bottom will prevent rocking and sliding.

DESIGNED BY

Randy Engels and Danny Ionazzi.

❧❧❧

Open-corner construction of parallel platforms is a standard construction method used in most professional scene shops. The method described below reduces the time needed to build parallel platforms, making them a practical platforming choice.

BUILD A JIG

Begin by taking two short pieces of the framing stock you intend to use and attach corner block material to their faces. Next, lay them out end to end and fasten the two pieces together with the type of hinge you intend to use. Bend the pieces to a 90° angle — this defines the open-corner measurement, which includes the gap created by the hinge, the corner-block thickness, and the framing-stock thickness. See Figure 1.

Each parallel is made up of seven framing members called "gates," five identical pieces for the width, and two for the length. See Figure 2. The length of each gate is determined by subtracting twice the open-corner measurement from the overall length of each gate.

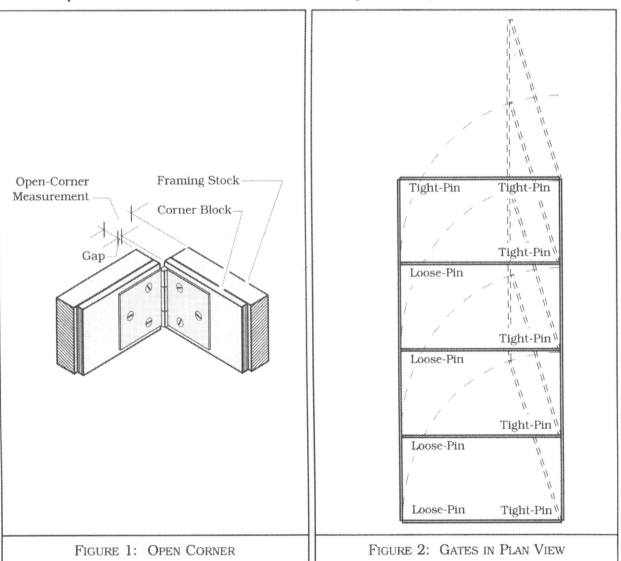

FIGURE 1: OPEN CORNER   FIGURE 2: GATES IN PLAN VIEW

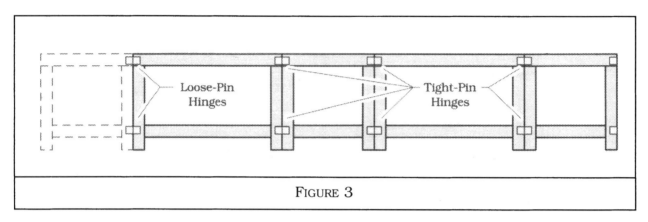

FIGURE 3

ATTACH THE HINGES

The real convenience of the open-corner construction method is that hinges need not be attached while the parallel is in the upright position. Lay out the two side pieces and two end pieces as shown in Figure 3 and hinge them together, attaching the loose-pin hinges first. Pull the pins and move the end pieces to the other side, and then attach the tight-pins. Next, place hinges on the pieces that are to serve as internal members, remembering to "hang the knuckles" of the hinges so that their width will correspond to the end gates. They can now be hinged in the appropriate places to the side gates still laid out on the bench, attaching the loose-pin hinges first, removing the pins, and then attaching the tight-pin hinges. When all the hinges have been attached, pick up the unit, fold it to the correct shape, and make the remaining hinges. Figure 2 will be helpful in determining the location of loose-pin and tight-pin hinges.

ADDITIONAL NOTES

Parallels are most useful when raised platforming needs to be between 18" to 6'-0" tall. They provide a lasting, easily stored platform system. To increase their durability, glue the joints and bolt one hole on each side of every hinge. Using corrugated fasteners on joints opposite corner blocks is also a good idea. As lids, use sheets of $\frac{3}{4}$" plywood with blocks set back from the corners to hold them in place.

❧❧❧❧

A lap-joint deck is a solid squeak-free alternative to traditional bolted decks. Individual platforms are constructed with carrying strips and corresponding deck overhangs as shown in Figure 1, and are screwed together from above during load-in.

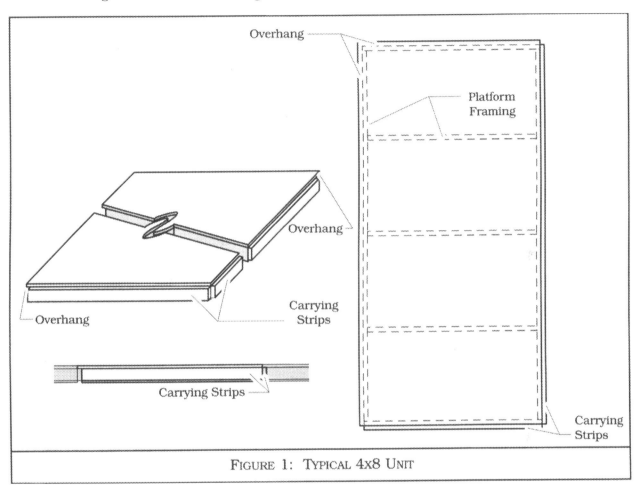

FIGURE 1: TYPICAL 4x8 UNIT

The frame and top may be constructed of the materials used for standard platforming. $\frac{5}{4}$x3 pine is a good choice for the carrying strips, although 2x stock may be substituted. The carrying strip must be firmly attached to the framing members. Gluing-and-nailing is suggested.

The plywood overhang should exceed the width of the carrying strip by $\frac{1}{16}$". For example, if the $\frac{5}{4}$ carrying strip measures $1\frac{1}{16}$", the overhang should be $1\frac{1}{8}$". This allows for variance in lumber dimensions and construction errors, thereby eliminating gaps between platform tops. Since the vertical surfaces of the framing do not bind between platforms, the deck is quieter.

PLANNING THE DECK

1.   All platforming is planned relative to the "key piece," usually positioned downstage center as shown in Figure 2. Other units are then laid out as with any other deck. Careful considera-tion must be given to the location of the overhangs and carrying strips. In Figure 2, overhangs are indicated with the letter "o." A carrying strip must be placed on any platform side that meets an "o" side. Note that the key piece would have carrying strips on three sides.

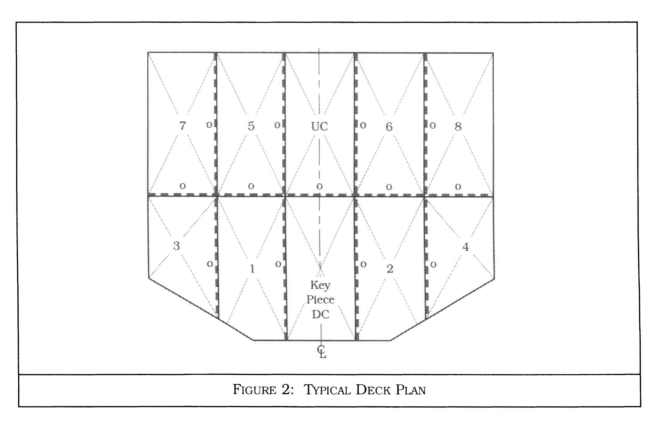

FIGURE 2: TYPICAL DECK PLAN

2. Platforms on one side of the key piece are usually labeled with even numbers, while units on the other side are labeled with odd numbers.

3. Obviously, it is not necessary to provide carrying strips or overhangs along the outside perimeter of the deck.

4. Tracks, turntables, and odd-shaped pieces may be planned into a lap-joint deck by applying the above procedures.

ASSEMBLY

Since the location and orientation of the key piece determine the placement of the entire deck, considerable care should be taken to place it precisely during load-in. Other units are then placed around the key piece and screwed through the overhang into the carrying strip. Three or four $1\frac{1}{2}$" flat head wood screws along each side will provide a sufficiently strong connection.

Many theatres stock a set of 4' x 8' platform units made with 2x4 framing and plywood tops. Though units like these can be built quickly and inexpensively and can be repeatedly assembled into decks by simply bolting on a new set of 2x4 legs, they are far from ideal. They frequently need repair, their assembly into decks often requires the use of shims, and the decks they form always seem to squeak.

The deck system described in this article has several advantages over its alternatives. Consisting of 4' x 8' stressed-skin panels and a minimal number of Styrofoam® supports secured to the stage floor, this system is more quickly and easily assembled, more rigid, and less likely to squeak than others.

THE STRESSED-SKIN PANELS

A typical 4' x 8' stressed-skin panel in this system uses a $\frac{5}{8}$" CD plugged and touch-sanded top skin and a $\frac{3}{8}$" plywood bottom skin on a frame consisting of five $\frac{3}{4}$" x 3" pine stringers and two $1\frac{1}{2}$" x 3" fir headers. Two Roto Locks® bolted onto the 8' side of each panel's top skin allow the panels to be connected firmly together during deck assembly. To avoid having to notch the stringers to accept the Roto Locks® and to provide a good handhold, the frame and bottom skin are made 4" narrower than the top skin. In addition, the bottom skin of each panel is fitted with four bolt plates like those in Figure 1 that allow for a firm connection between the panels and the Styrofoam® supports.

Complete design and construction details for stressed-skin panels appear in the pamphlet *Plywood Design Specifications, Supplement #3*, published by the American Plywood Association, 1119 A Street, Tacoma, WA 98401.

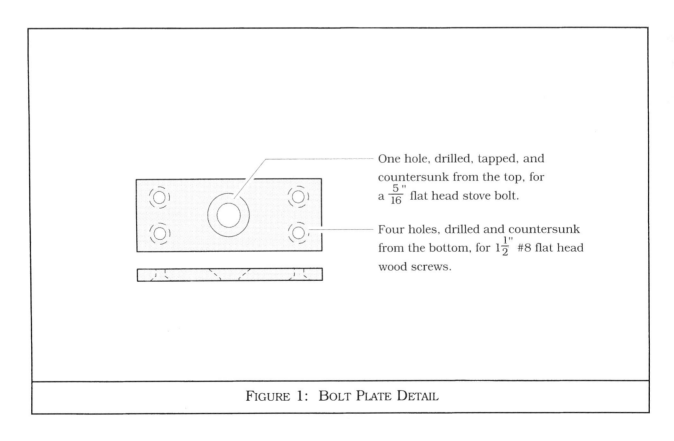

One hole, drilled, tapped, and countersunk from the top, for a $\frac{5}{16}$" flat head stove bolt.

Four holes, drilled and countersunk from the bottom, for $1\frac{1}{2}$" #8 flat head wood screws.

FIGURE 1: BOLT PLATE DETAIL

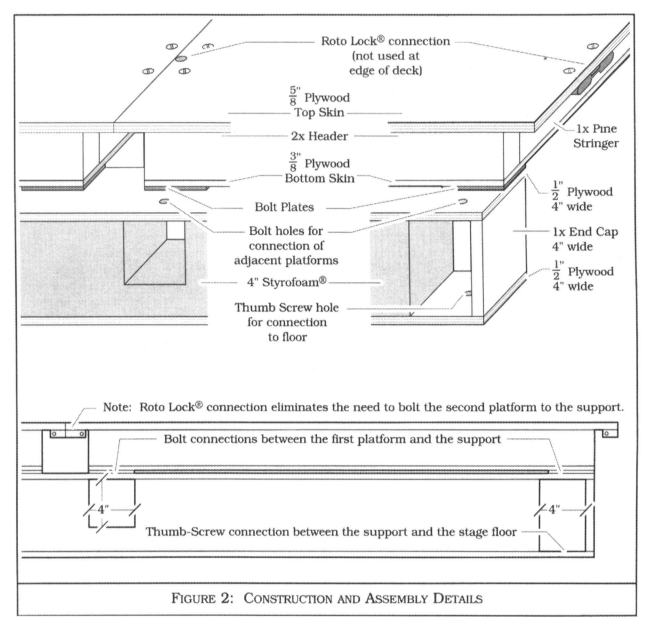

Roto Lock® connection
(not used at
edge of deck)

$\frac{5}{8}$" Plywood
Top Skin

2x Header

$\frac{3}{8}$" Plywood
Bottom Skin

Bolt Plates

Bolt holes for
connection of
adjacent platforms

4" Styrofoam®

Thumb Screw hole
for connection
to floor

1x Pine
Stringer

$\frac{1}{2}$" Plywood
4" wide

1x End Cap
4" wide

$\frac{1}{2}$" Plywood
4" wide

Note: Roto Lock® connection eliminates the need to bolt the second platform to the support.

Bolt connections between the first platform and the support

4"   4"

Thumb-Screw connection between the support and the stage floor

FIGURE 2: CONSTRUCTION AND ASSEMBLY DETAILS

THE STYROFOAM® SUPPORTS AND THEIR CONSTRUCTION

These supports have enough "give" to compensate for a reasonably uneven stage floor, and, despite their light weight, they are easily able to support the compressive loads placed on them.

1. Cut 4" Styrofoam® boards to a height 1" less than the finished support height. A raked deck can be accomplished either by building tapered supports to be laid out in the axis of rake or by beveling the top of the Styrofoam® and laying the finished supports out across the axis of the rake.

2. Notch the Styrofoam® wherever connections to a panel or the floor are needed. See Figure 2.

3. Glue $\frac{1}{2}$" x 4" plywood strips to the top and bottom of the Styrofoam® and cap the end of each support with a piece of $\frac{3}{4}$" x 4" pine cut to fit between the plywood strips. See Figure 2.

4.   Drill $\frac{1}{2}$" holes through the plywood strips wherever the support is to be attached to the stage floor or to a panel.

TYPICAL DECK ASSEMBLY

A 1'-high 16' x 16' deck would require eight panels and three 16'-long supports. The positive connections between panels, between deck and supports, and between supports and stage floor assure a firm playing surface. By following the sequence outlined here, a crew can quickly assemble the entire deck without having to crawl underneath to make connections.

1.   Place the Styrofoam® supports in position on the stage. The panels require support along their 4' sides only. Mark the location of the connections between the supports and the floor.

2.   Move the supports and set a threaded insert into the floor at each point of connection.

3.   Set the supports in place and connect them to the stage floor. Use thumb screws to speed deck assembly. If the deck is to be more than 1'-0" high, provide suitably spaced diagonal bracing between the supports.

4.   Set the end panel of each row in place on the supports and use wing nuts on the four bolts that connect the panels to the supports.

5.   Roto Lock® the next panel in each row to the first and bolt the accessible 8' side to the supports.

6.   Repeat the procedure until the deck is complete.

In addition to providing a firm playing surface, the 4"-thick stressed-skin panels stack very efficiently for storage. Moreover, their framing lasts longer than conventional 2x4 framing because it is not subjected to the wear and tear of repeated legging connections.

❧❧❧❧

Two Methods of Constructing Terrain Decks

David Cunningham

This article describes two methods of constructing decks with contoured surfaces. These decks rest on substructures consisting either of oddly raked platform units or of contour ribs affixed to gates (frames) or studwalls. Plywood lath woven or arranged in layers on top of the substructure smooths out the contours while providing the surface with the necessary strength. Successive layers of increasingly flexible materials such as chicken wire, carpet padding, and burlap give the deck its finish contours and texture.

The lath used can be cut in many widths and shapes from lauan or plywood $\frac{1}{8}$" to $\frac{1}{2}$" thick. However, the best material for lath is $\frac{1}{4}$" plywood ripped into 2" strips and cut on the bias, *i.e.*, diagonally across the sheet. Lath need not be applied over every square inch of the deck's surface: spacing at whole or half a strip's width is acceptable. Before the final coverings are applied, test the bearing strength of all areas by walking on them.

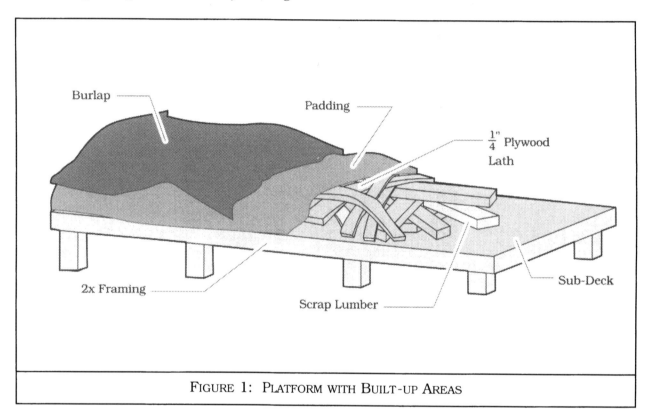

FIGURE 1: PLATFORM WITH BUILT-UP AREAS

THE PLATFORM-SUPPORTED SYSTEM

See Figure 1. For gently sloping terrain, construct a set of platforms (compound-raked as necessary) that roughs out the dominant slopes. Individual platforms need be neither rectilinear nor level along any edge. Whatever the shape of any platform, identify the three points of elevation that establish the plane of its surface. In determining the height of these controlling elevations, allow for the estimated thickness of the built-up lath, which can be interwoven and broken off as necessary to form the desired contours. Where a strip does not quite reach a nailing surface, subsequent strips can be woven beneath it to provide better bearing strength. At the same time, however, lath should not be forced to bend where it doesn't want to bend.

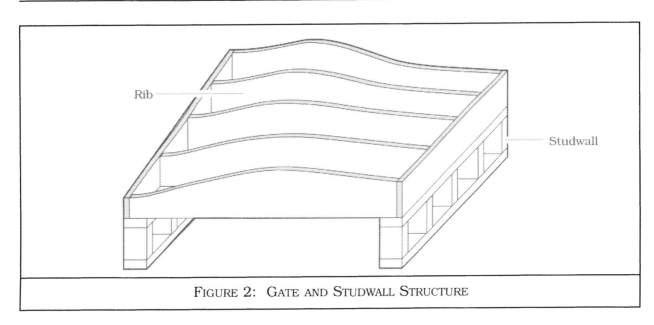

FIGURE 2: GATE AND STUDWALL STRUCTURE

THE RIB-SUPPORTED SYSTEM

For terrain with compound curves or abrupt changes in elevation, a series of plywood ribs braced together at one-foot intervals provides a more suitable framework than a set of platforms. The contours of some rib-supported decks may permit the installation of rib units on top of 2x4 studwalls as shown in Figure 2. If, however, the contours of a ribbed deck are such that the ribs cannot be parallel to each other, the ribs can be built as the upper rails of individual 1x gates.

Whether rib units are supported by gates or studwalls, the location of the breaklines between them should be based on both handling considerations and abrupt changes in grade. Contours along the perimeters of adjacent rib units should be eased into each other by making any necessary adjustments in elevation. Finally, the entire perimeter of each rib unit should be framed, as shown in Figure 2, to provide a nailing surface at each seam. In a rib-supported system, the lath is layered rather than woven. A first layer of lath is applied as close to perpendicular to the ribs as possible. A second layer is then attached diagonally across the first; and a third layer, parallel to the ribs.

ATTACHING THE LATH

Air-driven fasteners reinforced with glue produce the best connection between the lath and its substructure. This approach is faster than the use of nails or screws. Moreover, nailing by hand tends to shock previous connections loose, and screws too often split the wood.

FINAL NOTES

Doing a trial setup of the substructure for terrain decks before applying the lath is advantageous. Set the substructure units in place separated from each other by the width of a saw blade. Then, apply lath to the entire deck, smoothing the transition between units. Finally, cut the breaklines into the lath with a reciprocating saw.

A Non-Skid Groundcloth

Philip E. Hacker

The following method for making a non-skid groundcloth was devised at the University of New Mexico to meet the needs of a specific touring production in which cast members stilt-danced on the groundcloth.

The groundcloth had to be tacky enough to provide a sure footing for the stilts strapped to the actors' legs. It also had to resist bunching up around the base of the stilts, and it had to be kept from sliding across the surface of the deck.

MATERIALS AND EQUIPMENT

Large bucket	Paint brushes
8oz Duck or scenery cloth	Deck broom
Phlex-glue	Fairly coarse sawdust
Phlex-glue Super Plastisizer	Sizewater
Latex paint	6" or 8" Sonotube®

Phlex-glue and Phlex-glue Super Plastisizer are produced by Spectra-Dynamics of Albuquerque, New Mexico.

PROCESS

Lay out the groundcloth, leaving it oversized by 2' along each edge. Using heavy cotton thread, machine-sew the fabric strips together with flat-felled or other appropriate seams.

Attach the groundcloth seams-up on a paint deck or seams-out on a paint frame. Staple the fabric around all edges, size it, and allow it to dry.

To prepare the backing mixture, mix the Phlex-glue, latex paint, and Plastisizer in a ratio of 1 gallon to 1 gallon to 2 ounces. The color of the latex paint used in the backing mixture should be compatible with the color scheme of the groundcloth's top surface.

Apply a first coat of the backing mixture to the cloth with a brush, using the cross-hatch method to work the mixture into the nap. Although the backing mixture will never cure completely, it is dry enough to work with if, when grasped between thumb and forefinger, it does not slide across the surface of the fabric. Generally, a second coat of the mixture can be applied after about three hours' drying time. Allow the second coat to dry for about five hours.

If the groundcloth has been treated on a paint frame, carefully remove it and lay it out on the floor, coated side up. Avoid letting the coated surface fold back on itself: only with great difficulty can accidentally folded fabric be unfolded. Once the groundcloth is in place seams-up, scatter a generous amount of sawdust along one edge. Then, having removed your shoes, sweep the sawdust across and off the coated fabric with a stiff deck broom. It is not the coarse sawdust, but rather the fine particulates remaining on the fabric that make the groundcloth non-skid. Flip the groundcloth over and slide it back and forth across the deck to insure even distribution of the fine particulates.

Next paint the top side of the groundcloth. The author recommends the use of diluted latex paints to provide the desired skid resistance and to mask any bleed-through of the Phlex-glue mixture. When painting is complete, trim the groundcloth to its final size.

STORAGE AND TRANSPORTATION, INSTALLATION, AND STRIKE

The groundcloth should be stored and transported by rolling it evenly around a 6" or 8" Sonotube®. The paraffin-impregnated surface of the tube will keep the material from adhering to it.

To install, roll the groundcloth out onto the stage and place it in position. Turn on all available stage lights to heat and soften the undercoating and to flatten any wrinkles. Then, tape down the edges with cloth tape.

During strike, allow the surface to cool before removing tape in order to prevent stretching the edges.

❧❧❧❧

A Laminated Plywood Turntable

Thomas P. Sullivan

Traditional turntable designs cause some difficult problems for technical directors. Accuracy in making the beveled cuts required for their framing is time-consuming and chancy, and the resulting joints often creak and groan. In addition, because the casters move across the floor below, running power or control cables to the pivot is difficult; and debris and dirt on the floor itself make the turntable's movement bumpy and noisy.

One way to address these problems is to eliminate their causes. A frameless turntable can be built more quickly, for instance, and if the casters are attached, inverted, to the floor, nothing under the turntable can interfere with their smooth and quiet operation. This article describes one such turntable system.

GENERAL DESCRIPTION

Figure 1 clarifies the basic design of this system, which comprises four major components: the turntable, the bed of inverted casters, the pivot, and a drive mechanism. Of these four, the pivot and the drive mechanism resemble those used in familiarly framed turntables and will not be discussed here.

The caster bed comprises two rings of fixed casters, inverted and bolted to a plywood panel, their axles aligned with the radii of the turntable. See Figure 1A. The most complicated component of the system is the turntable, a three-layer lamination of plywood panels. See Figure 1B. The turntable's construction will be outlined later in this article.

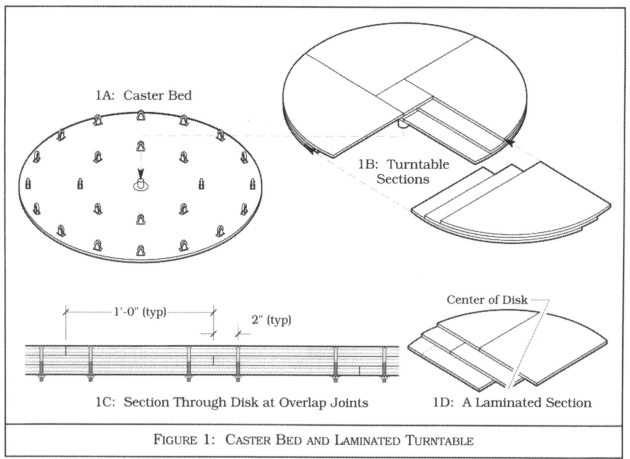

1A: Caster Bed

1B: Turntable Sections

1'-0" (typ)   2" (typ)

1C: Section Through Disk at Overlap Joints

Center of Disk

1D: A Laminated Section

FIGURE 1: CASTER BED AND LAMINATED TURNTABLE

DESIGN VARIABLES

Three interdependent elements need to be considered in designing turntables based on this system: the load to be carried by the turntable, the distance between casters, and the size of the sections of the turntable disc. The thickness of the turntable is based on projected loading conditions and structural design information provided in the American Plywood Association's pamphlet *Plywood Design Specification.* The distance between casters and caster rings is determined by the type and thickness of plywood panel chosen. The size and number of sections in the turntable disc is determined by weight, handling, and storage considerations. The example in Figure 1 is designed with four wedge-shaped sections.

DESIGN CONSTANTS

In order to take full advantage of plywood's strengths, a turntable based on this model should be built with the following notes in mind:

1. Each layer of plywood in a section should be laid out with its face grain running parallel to the face grain of the other layers.

2. The face grain in each section should run either parallel or perpendicular to the radius of the turntable as much as possible.

3. The seams between the pieces of plywood in each section should be staggered.

4. The joints between sections and between the sections and the pivot are intrinsically the weakest parts of the turntable disc.

5. The overlap joint should be as wide as practicable. The sections are joined by bolts spaced 6" apart o.c. in rows 2" from either side of each joint as in Figure 1C. T-nuts epoxied into the bottom of the disc will speed assembly.

MODEL SPECIFICATIONS

The 12'-diameter model in Figure 1 is designed to the following specifications:

LOAD	50 psf (a fairly typical deck rating).
PANEL GRADE & THICKNESS	$\frac{3}{8}$" CD interior sheathing.
CASTER SPACING	34" between the pivot and the first caster ring and between the first and second caster ring.
JOINT OVERLAP	1 foot. See Figure 1C.

CONSTRUCTION NOTES

1. Spread an even coat of wood glue over the area of contact between the panels in each section, but do not put glue on joint surfaces of adjacent sections.

2. Once all of the panels in a section are in place, nail through the layers in a 6" grid pattern. This will keep the necessary pressure on the plywood until the glue has dried.

3. Once the sections have dried, mark and cut the desired circumference of the turntable.

4. Once the sections have been completed, assemble them to form the entire disc.

A Plasticene/Styrofoam® Deck Plug

David C. Perlman

THE PROBLEM

The action for the Yale Repertory Theatre's production of Edit Villarreal's *Crazy from the Heart* involved plunging narrow wooden stakes into the ground during a Native American gambling game. During each performance, a total of thirty-two sticks were to be repeatedly stabbed into the deck, making a total of over 700 holes throughout rehearsals and the run. Since the Masonite®-covered stressed-skin-panel deck had to work for both this show and another with which it played in rotating repertory, the deck had to retain a uniformly smooth, planar surface. Thus, for the gambling scene, we had to devise a plug that met several criteria:

1.   The surface had to resemble that of the surrounding deck in color and texture.

2.   The contents had to make the stabbing look and sound plausible.

3.   The stakes jammed into it during the performance had to remain upright.

4.   It had to cover a reasonably large area to accommodate the action of the play, yet could not be overly awkward to handle.

5.   The area where the game took place would have to look "clean" each night, the previous performance's indentations invisible.

We ruled out using sand alone as the filler for several reasons. Traffic across sand would have left footprints, making the plug surface visually different from that of the surrounding deck. Sand would also have made the plug too cumbersome for easy repping, and would have offered the obvious potential for creating a mess if the plug were dropped. Worst, so shallow a sand bed would not provide enough grip to keep the stakes vertical. We also ruled out a completely Styrofoam® filler because it would squeak when stabbed and because the cost prohibited replacing it after each use.

THE SOLUTION

The plug we built was a 1x3-and-plywood box measuring 36" x 40" x $2\frac{3}{8}$" deep. After building the box, we glued a layer of 1" Styrofoam® to the inside with Fastbond 30®. We then applied a coating of Fastbond 30® to the Styrofoam® and laid in $\frac{1}{2}$"-thick slabs of plasticene. We topped the assembly off with a layer of sand — not deep enough to result in footprints, but deep enough that the audience could see sand thrown into the air during the game.

This idea worked extremely well. The key to its success was the plasticene, which kept the stakes vertical, muted the sound of the Styrofoam®, and made the action quite realistic. The most valuable aspect of the plasticene was that it could be remolded to fill in the holes that the stakes made.

NOTES ON CONSTRUCTION

Commercially available plasticene comes in grades numbered 1 through 4 — soft to very firm. Two-pound bricks of Grade 2 worked well for this application. Wrapping the bricks in plastic bags and submerging them in hot water for about ten minutes made the material very pliable. For an even lower-cost approach, you might make your own plasticene following the instructions in the *Technical Brief* article, "High-Volume, Low-Cost Modeling Clay" by Jon Lagerquist.

The design for a Yale Repertory Theatre production of *Scapin* called for the main portion of the deck to be a 22'-diameter circle, 8" above deck height. In addition to the expected qualities of quietness and durability, it had to be soft enough for acrobatics and pratfalls, and it had to have a non-skid surface that would take paint well.

The first objective was to find the right resilient material. The first material researched, $\frac{1}{2}$" industrial felt, was entirely too costly and not nearly soft enough. The next possibility, a custom-made wrestling mat, would have cost $1,400.00 and taken four to six weeks to arrive. The material we finally chose was Ethafoam®, a trade name for a group of closed-cell polyethylene products. Many are familiar with it in the rod form that is split to make moulding or bumpers for scenic pieces. For this application, we purchased rolls of sheet Ethafoam® $\frac{1}{2}$" thick by 4'-0" wide by 125' long. The technical name for the product is "closed-cell polyethylene, density range 1.6 to 1.9 pounds per cubic foot." According to the vendor, this description will get you the same product nationwide. The cost of each 500-square-foot roll was $210.00, and for this show we bought two rolls.

Having chosen a material, our next task was to attach it to the floor and to attach a groundcloth to it as a paint surface. Attaching the Ethafoam® to the deck became a non-issue for us because the deck construction design we developed included a plywood ring that would hug the perimeter of the disc tightly and hold the Ethafoam® in place. For *Scapin*, we used two layers of Ethafoam® sheets, gaffers' taping the seams of each layer's adjacent panels together and then installing the two layers with their seams perpendicular to each other as shown in the partial plan in Figure 1.

To attach the groundcloth to the Ethafoam®, we used 3M contact adhesive NF-30®, familiarly known as "green glue." Knowing that we wouldn't be able to slide the groundcloth into place on the Ethafoam® once the green glue had been applied, we approached the task as though we were laying out a giant pie crust. First, we laid out the Ethafoam® and placed the untreated groundcloth on top of it, stretching the groundcloth smooth and making sure that its painted pattern was oriented correctly. Next we folded the groundcloth back on itself to expose half of the Ethafoam® and the back of the corresponding half of the groundcloth. Once the green glue had been applied and become tacky, we used a 24' roll tube as a sort of oversized rolling pin to roll the groundcloth into place. Finally, we stapled the edge of the groundcloth to the perimeter of the disc and entrapped the entire assembly with the ring as shown in Figure 2.

We were very happy with the results of this approach: the bond between the groundcloth and the Ethafoam® proved highly effective against actor traffic; the elaborately painted soft deck held up well over the course of a four-week run; and we recovered whole sheets of reusable Ethafoam® at strike, discovering that the green glue would peel away with little effort.

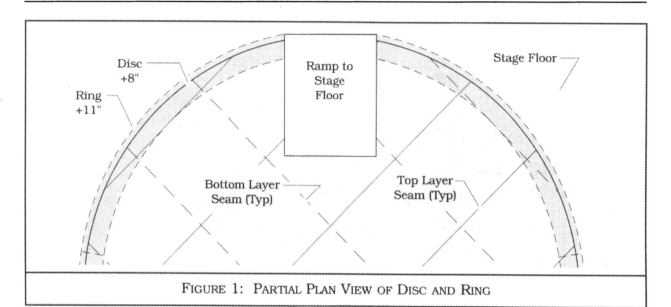

FIGURE 1: PARTIAL PLAN VIEW OF DISC AND RING

FIGURE 2: SECTION THROUGH DISC AND RING

Scenery Electronics

Sliding Electrical Contacts
Arthur Oliner

Sliding electrical contacts provide an easy way to get power to electrically operated devices mounted on tracked pallets or free-rolling wagons. The use of such contacts, which can carry either audio signals or electrical currents, eliminates the need for cable that must be paid in and out as the pallet or wagon moves. Sliding electrical contacts are easily constructed in the shop.

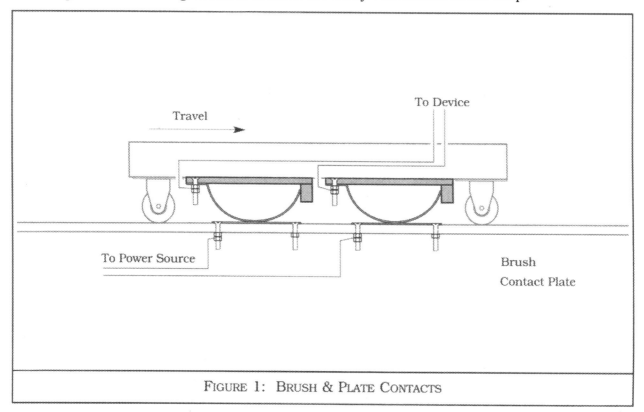

FIGURE 1: BRUSH & PLATE CONTACTS

OPERATION

The contact consists of two components: a flexible brush, and a contact plate. The brush, a curved piece of copper or brass, is mounted on the underside of a wagon to bear on the floor with some force. The contact plate, also copper or brass, is floor-mounted at the wagon's playing position. Power is supplied to the wagon by the contact of the brush and the plate. See Figures 1 and 2. One sliding contact is needed for each conductor.

DESIGN PARAMETERS

Contacts must be designed to provide sufficient contact area to carry the current needed to operate the device. To determine the necessary contact area, first determine the amount of current needed to operate the device. Then refer to an electrical handbook to discover what gauge wire would be needed to carry that current. This data appears under the heading "ampacity." The same handbook will reveal the cross-sectional area (in square inches) of that wire. That cross-sectional area is the necessary surface area that must remain in contact between the two components.

According to the National Electrical Code, exposed contacts may not carry more than 50 volts. The platforms' playing position prevents exposure of the live contact plate. When the platform is moved, however, current must be disconnected from the contact. Disconnection can be assured by the use

Arthur Oliner   Sliding Electrical Contacts

of a dual safety mechanism consisting of one switch that is triggered mechanically as the platform moves from its playing position and a second switch that can be activated by an operator at a remote station.

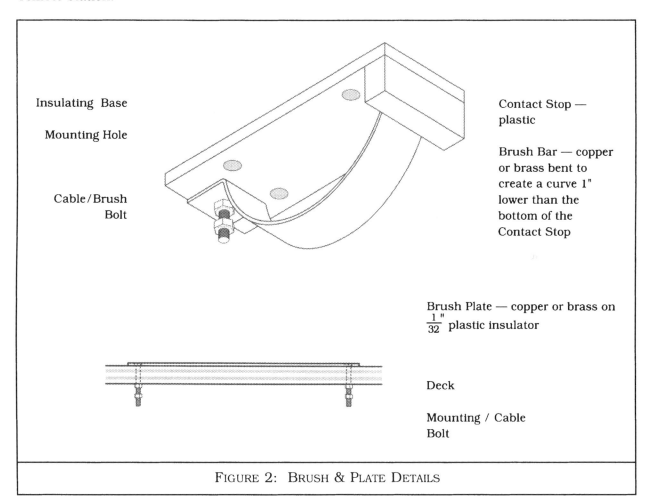

Insulating Base

Mounting Hole

Cable/Brush
Bolt

Contact Stop —
plastic

Brush Bar — copper
or brass bent to
create a curve 1"
lower than the
bottom of the
Contact Stop

Brush Plate — copper or brass on
$\frac{1}{32}$" plastic insulator

Deck

Mounting / Cable
Bolt

FIGURE 2: BRUSH & PLATE DETAILS

CAUTIONS

1. Check local and national electrical codes to insure proper installation and operation of the contact.

2. Make the contact plate long enough to allow for variation in the travel of mechanically operated pallets or wagons.

3. Maintain proper insulation between the contact and surrounding materials to prevent accidental electrification of parts not intentionally included in the electrical system.

173

A Low-Voltage Remote Controller for Special Effects *Kenneth J. Lewis*

Special effects that respond to a simple on/off control signal can be operated by a Magnetic Reed-Switch (MRS) Controller. The MRS Controller will operate such effects as "wheat" lamps, star-drop circuits, bells, buzzers, and solenoids without requiring sophisticated electronics and construction.

An MRS operates by reacting mechanically to the presence of a magnetic field. It is widely and cheaply available because of its use in intruder-alarm systems. In such systems, the switch is placed on a door jamb, and the magnet on the door. When the door is closed, the magnet holds the switch closed, completing a normally closed circuit. When the door is opened, the magnet moves away from the switch, breaking the circuit.

In a special effects controller, any number of MRSs are attached to a plywood base in any desired pattern. One terminal of each MRS is wired to the power supply; the other, to a separate wire for each special effects device. The neutral wires of all of the devices can be ganged and are returned to the power supply. See Figure 1.

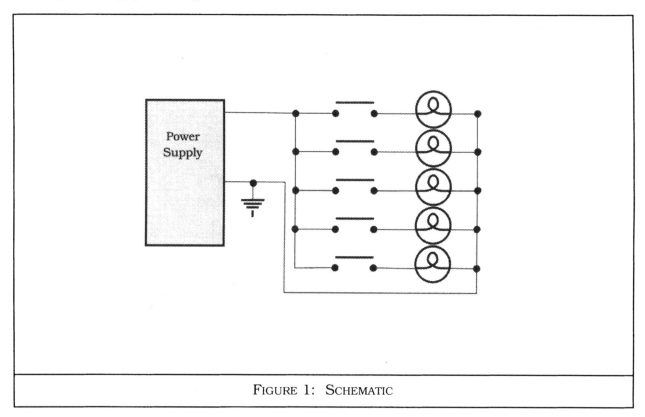

FIGURE 1: SCHEMATIC

If sides are attached to the plywood base as in Figure 2, they can support a thin sheet of Plexiglas® on which a magnet can rest in close proximity to the switches. The space between the Plexiglas® and the MRSs is determined by the strength of the magnet and the sensitivity of the switches used. The spacing can easily be determined through trial.

The operator of the MRS Controller can quickly set and adjust cues. Recording cues in the form of a map drawn on the Plexiglas® facilitates repetition of cues.

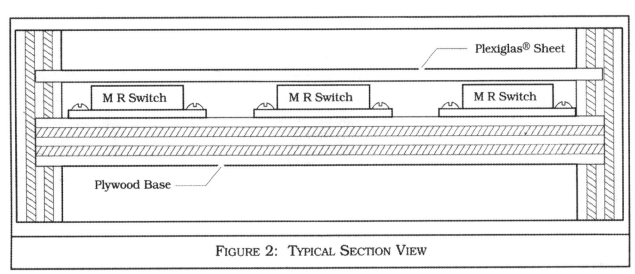

FIGURE 2: TYPICAL SECTION VIEW

The MRS Controller is a simple stage effects tool. It can be built easily and adapted to a variety of circumstances. For example, it has been used to operate a series of small lamps placed about the scenery for a production of *Peter Pan*. In that production, the sequential lighting of the lamps represented Tinkerbelle's location and movement.

Further, the Controller's simple nature gives it very high reliability and allows cueing problems to be solved quickly. Through its use, elaborate special effects that can delight an audience can be easily reproduced.

¿●¿●¿●

Though LEDs (light-emitting diodes) are commonly used as everything from panel indicator lamps to the tiny flashers on sun visors, many technicians continue to find them somewhat mysterious. This article describing two simple LED circuits should remove some of that mystery and hint at the broad range of possible theatrical applications.

The constant-on LED circuit represented in Figures 1 and 2 could be used in place of a standard 120V cuelight system, freeing up house circuits for other uses. Similarly, it could be used to provide necessary escape-platform lighting or to replace glow tape in some other applications.

The flashing-LED circuit depicted in Figure 3 might be used to indicate the location of offstage emergency equipment such as fire extinguishers and first aid kits. Its low voltage and small size also recommend it for use in some self-contained onstage special effects such as flashing lights on costumes.

CONSTANT-ON LED CIRCUIT

In this simple series-wired DC circuit, both the size of the battery and the number of LEDs used are critical factors. The 9V battery will run four LEDs at normal brightness for a reasonably long period of time. While a 9V battery cannot power more than four series-wired LEDs, it would quickly fry a single LED. A constant-on LED circuit uses the following parts:

> 9V Battery
> 9V-Battery Cap
> Package of Assorted LEDs
> Single-Pole, Single-Throw (SPST) Toggle Switch
> A Length of 22-gauge Wire

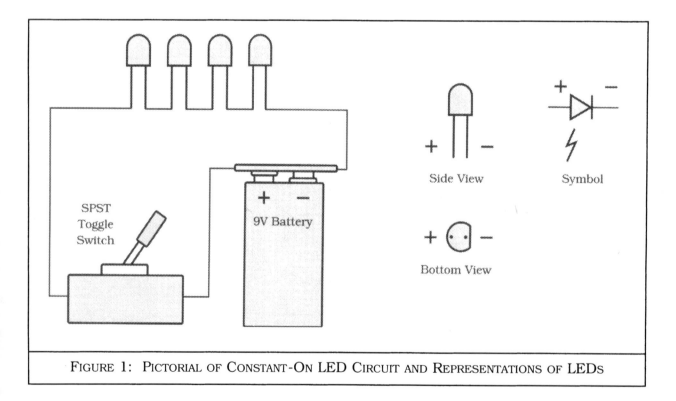

FIGURE 1: PICTORIAL OF CONSTANT-ON LED CIRCUIT AND REPRESENTATIONS OF LEDS

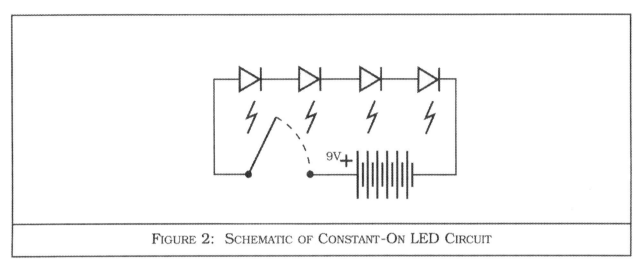

FIGURE 2: SCHEMATIC OF CONSTANT-ON LED CIRCUIT

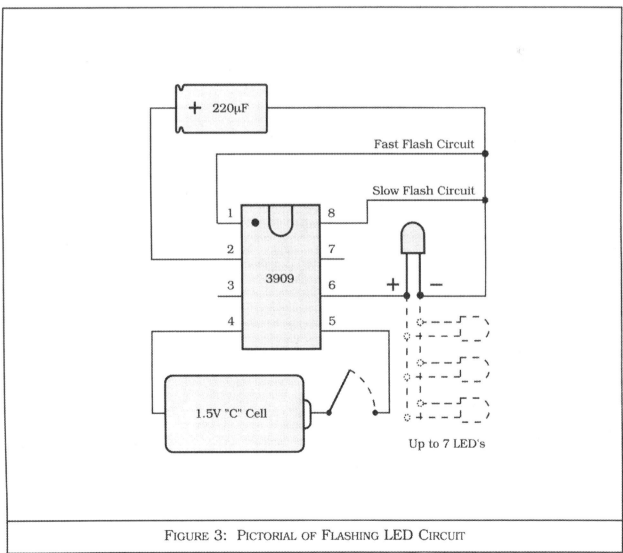

FIGURE 3: PICTORIAL OF FLASHING LED CIRCUIT

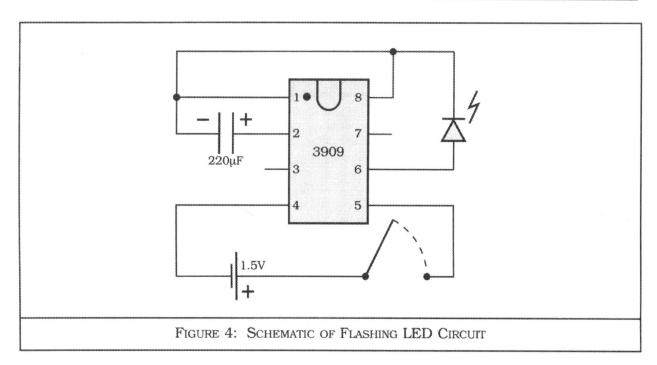

FIGURE 4: SCHEMATIC OF FLASHING LED CIRCUIT

FLASHING LED CIRCUIT

Though the integrated-circuit chip makes it somewhat more sophisticated, this is really a simple timer circuit that can operate one to seven parallel-wired LEDs. The components can be wired to produce different rates of automatic flashes. For a flash rate of approximately one flash per second, wire as shown. A slower flash rate (about one flash every three seconds) can be obtained by eliminating the wire to pin 1. The flashing LED circuit uses the following parts:

1.5V C Cell	LM3909 Integrated Circuit (Chip)
C-Cell Holder	Chip Socket
220µF Capacitor	Small SPST Toggle Switch
Package of Assorted LEDs	A Length of 22-gauge Wire

NOTES

1. All of the necessary parts for these and most other LED circuits can be found at any Radio Shack or other electronic supply store.

2. Since most of the components used in these and similar circuits are polarized, the circuits won't work if components are wired "backwards." During soldering, be sure to use appropriate heat sinks to avoid "cooking" the parts.

3. Clear and readable explanations of the basic principles that underlie the operation of these and many other electronic circuits can be found in such books as Horowitz and Hill's *The Art of Electronics* and Diefenderfer's *The Principles of Electronic Instrumentation.*

SCENARIO

You're the master electrician on a production that involves the use of many special effects, including smoke, lights, horns, bells, and whistles. The director insists that all effects be controlled by the Stage Manager, who is located 200 feet from the stage. Your heart sinks at the thought of running separate control wires for each effect. What do you do?

SOLUTION

As long as there can be a split second between the execution of any two effects' cues, the solution to your problem may be a Touch-Tone Relay Controller. Through only a singe pair of wires, this controller can activate up to sixteen different devices. The schematic of the system is illustrated in Figure 1.

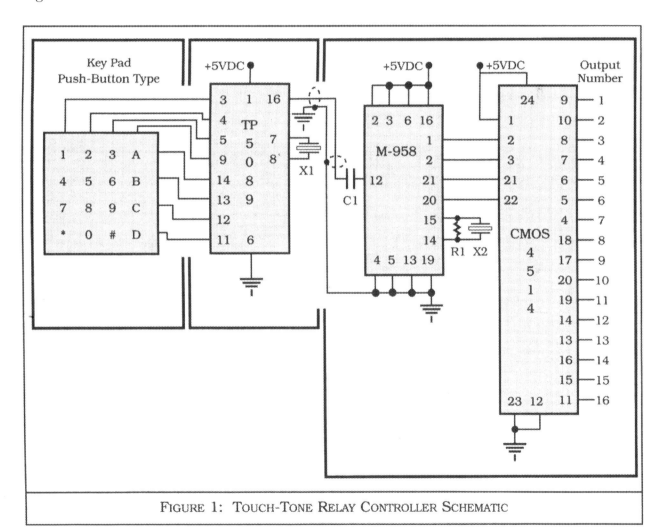

FIGURE 1: TOUCH-TONE RELAY CONTROLLER SCHEMATIC

A Touch-Tone Relay Controller for Special Effects

Steven E. Monsey

PARTS LIST

All components are commonly available at local electronics supply stores.

> TP5089
> Teltone M-958
> CMOS 4514
> X1, X2:   3.58 MHz Crystal (color-burst type)
> C1:   0.01 μF Capacitor
> R1:   1 MΩ Resistor

SYSTEM OPERATION

The operator pushes a single key on a telephone keypad, sending a standard touch-tone signal backstage where it is received and processed by a series of three integrated-circuit chips. The result of the processing is an electronic signal large enough to trigger a transistor and activate a relay-controlled effect. Because the range of relay and transistor types required for different effects is so wide, this article will not address the means of selecting them.

SYSTEM COMPONENTS

THE KEYPAD

Keypads can be purchased in twelve-key and sixteen-key models. The former is more readily available, but the latter can control four more effects since each cue is executed by pushing a single key. Sixteen-key keypads are available through electronics mail order suppliers and at electronics surplus stores. The author recommends that, whatever its number of keys, the keypad be powered by an independent power supply like the ones used for calculators.

THE INTEGRATED CIRCUIT CHIPS

1. TP5089 (The touch-tone encoder). This chip translates the touch-tone signal into audio signal.

2. M-958 (The touch-tone decoder). This chip receives the audio signal, which it converts into binary code.

3. 4514 (the 4-bit to 16-bit decoder). This chip receives and converts the binary code signal into a 25-milliamp, 0.5-volt signal capable of triggering transistors.

ASSEMBLY AND CONSTRUCTION NOTES

The device can be easily assembled, but builders should provide both the keypad and the decoder chips with sturdy housings to prevent against accidental damage. Finally, fitting both the keypad housing and relays to be driven by this controller with plug-type connections will facilitate transportation, installation, and interconnection with a wide number of effects.

❧❧❧

A Yale Repertory Theatre production of *Kiss of the Spider Woman* had a large, 2500-pound wagon that needed to be automated. Our usual drive method — a floor-mounted winch frame and a maze of muling blocks — was ruled out because of the shortness of the load-in time in touring. To meet these constraints, we built a unit whose self-contained drive consisted of a speed-controlled motor and a horizontal loop of roller chain, and whose stop points would be set by adjustable-position limit switches mounted in slots milled into the unit's frame.

In place of a system of relays, we decided to use our Programmable Logic Controller (PLC) to control this system. Even though this particular application is almost too simple for a PLC, its use saved us assembly and troubleshooting time, and it offered many features a relay controller lacked, *e.g.*, multiple operating modes and safety interlocks.

WHAT IS A PROGRAMMABLE LOGIC CONTROLLER?

A PLC is essentially a low-cost, heavy-duty industrial computer with pre-wired optically isolated inputs and either relay, transistor, or solid-state relay outputs. Originally designed to substitute for relay-logic control systems such as those found in elevator control panels and other large machinery, PLCs have lately become remarkably sophisticated. They have been used extensively in automated factories for some time and, because they are re-programmable, have more recently been used in Broadway shows. A system useful for all but the most demanding theatrical applications can be had for around $600, and yet, because of its re-programmability, the same system could be used to control motorized wagons, turntables, elevators, special effects, or — with the addition of electrically controlled valves — pneumatic or hydraulic devices.

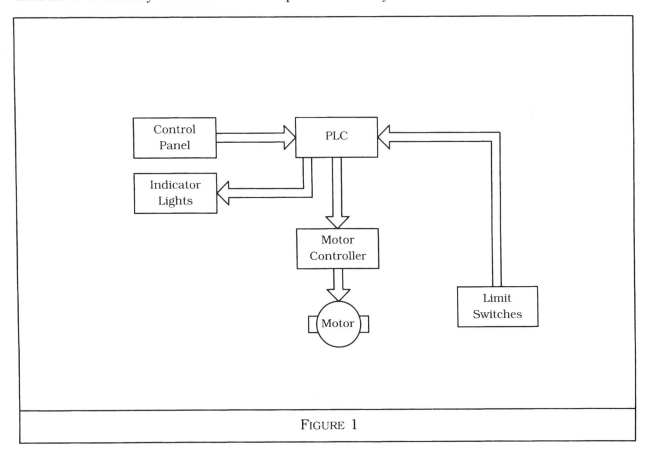

FIGURE 1

HOW DO THEY WORK?

In a typical application, the inputs of the PLC are connected to control or limit switches, and the outputs are connected to motors, solenoid valves, contactors, or other devices that need to be controlled by the computer. The PLC reads the condition of the inputs, and the internal program evaluates and executes the output states.

HOW ARE THEY PROGRAMMED?

The program for each application is written in ladder logic, a sort of sequentially executed relay logic schematic. Graphically, the program looks something like a ladder, hence the name. Programming is done using various types of "elements" that are analogous to relay contacts and coils. The program is stored in battery-backed RAM in the PLC, or in a detachable non-volatile ROM module. The program can be entered into the PLC either through the use of a small programming panel or, with the addition of special software, through an interface with a PC. The PLC itself possesses no means for entering a program, so one or the other of the above interfaces must be used.

WHAT APPLICATIONS ARE POSSIBLE?

PLC applications are essentially limited only by the programmer's imagination and budget. In the Broadway production of *Phantom of the Opera*, a single assembly-line-grade Allen-Bradley PLC controls 54 automated scenic units. This system has a custom full-color cuing interface and several levels of control redundancy and backup, but the heart of the system is the PLC running a program in ladder logic. The road companies of *Les Miserables* similarly used PLC systems to control large turntables. The Metropolitan Opera installed a large PLC to control their permanent turntable.

WHAT ARE THE WEAKNESSES OF A SMALL PLC?

Cueing input and sequential go-button operation like that of a lighting console is difficult to implement in a small PLC-based system. This is true primarily because, even though the units are designed to perform sequential tasks, the sequence of those tasks must be set in the PLC's internal program and, as such, is not easily changed by a non-programmer. This facet of the PLC's design would also make it difficult for an operator to skip forward or backward through cues if something unexpected happened during a performance. The smaller PLCs are more suited to operation where, for each cue, the operator selects a target position from all of the positions available for the show, and sends the scenic unit to that destination. Since many theatrical applications require only two destinations (*e.g.*, the up and down limits for an elevator), the PLC's design is appropriate, especially since it allows for sequences such as decelerating and creeping into selected target positions. While the ladder logic used by PLCs is very powerful, it is a low-level language and not intuitively understood. First-time users will need to spend time in experimenting with the language, while those who have previously dealt with relay logic circuits will find the transition to PLCs less difficult. Obtaining one of the PC-interface programs will also ease the learning process because these programs are self-documented and provide immediate graphic representation.

HOW IS A PLC BETTER THAN STRAIGHT RELAY LOGIC?

Given the tight schedules of the entertainment industry, the primary advantage in using a PLC is that the PLC simplifies the development and troubleshooting of control systems. Older systems — a maze of wires and relays — can be replaced by easily annotated and understood software. Consequently, programs can be changed with a few button presses rather than with a pair of wire strippers and a screwdriver. Further, since the "internal relays" exist only in the digital realm of the computer, there are no mechanical contacts to wear out. Expensive time-delay relays and other devices are emulated in software and, therefore, are much more cheaply implemented in a PLC. Since the primary circuitry of a PLC-based system exists in software, the PLC may be reused simply by writing a new program for each application.

The newer small PLCs are also amazingly sophisticated, with off-the-shelf fiber optic communication links, advanced digital-math capabilities, high-speed counters, analog I/Os, and even devices such as Mitsubishi's "Positioning Control Module." This PLC add-on can completely handle all the intricacies of a singe-axis of motion control including relative quadrature encoder feedback, acceleration, and deceleration.

WHERE CAN I GET ONE?

Allen-Bradley, Mitsubishi, Omron, and many other companies manufacture lines of PLCs. Howman Controls, a Mitsubishi distributor, also markets their own IBM PC interface program for Mitsubishi PLCs. Grainger sells small Omron PLCs, calling them "Sensor Controllers."

Allen-Bradley	Mitsubishi Electric Sales America
Industrial Control Group	Industrial Automation Division
1201 South Second Street	800 Berman Court
Milwaukee, WI 53204	Mt. Prospect, IL 60056
Howman Controls	W. Grainger Inc.
12 Garden Street	5959 West Howard Street
Edison, NJ 08817	Chicago, IL 60648

SOURCES

Associates and Ferren introduced me to PLCs. Steven E. Monsey provided information about *Phantom of the Opera* and *Les Miserables*. Geoff Webb contributed information about the Metropolitan Opera's turntable. The roller-chain drive for the Yale Repertory Theatre's application was devised by Alan Hendrickson.

Scenery Hardware

Floor-Mount Curved Track Using Polyethylene Pipe   *Mark Sullivan*

Occasionally, a scenic unit must track along a curved or irregular path. Construction of the track for such a situation can be expensive, complicated, and time-consuming. For situations in which the moving unit is relatively light, however, split polyethylene pipe can be used as a quick and inexpensive track. Polyethylene pipe is available through hardware stores and plumbing supply outlets in lengths up to 100', allowing for a seamless track. It is noted for its strength and high wear-resistance, making it durable, quiet, and maintenance free. Finally, the pipe can be worked using conventional hand and power tools. For tracking applications, 1"-diameter Type 1 pipe is best. The low ratio of wall thickness to pipe diameter provides a good balance between rigidity and flexibility.

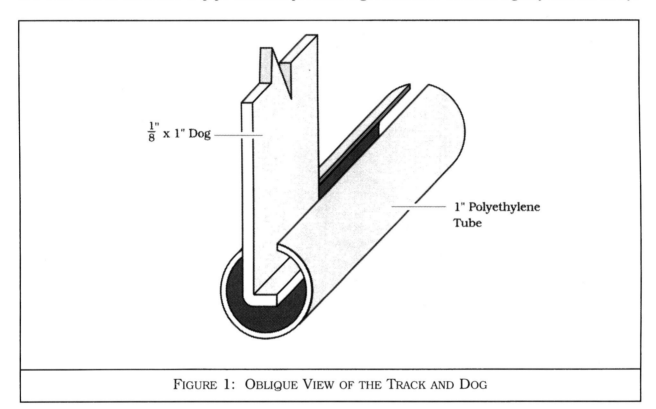

$\frac{1}{8}$" x 1" Dog

1" Polyethylene Tube

FIGURE 1: OBLIQUE VIEW OF THE TRACK AND DOG

SPLITTING THE PIPE

Cut the pipe to length, adding 2' for waste because the pipe curls at the ends making accurate splitting difficult. Use a table saw to split the pipe along one side so that the cross section is a "C" shape. Construct a sturdy jig to hold the pipe firmly so the blade cuts a straight path down the center of the pipe. Use a fine-toothed blade to do the cutting. This will assist in making a quiet and smooth running surface for the guides. While cutting, keep the pipe from rotating in the jig.

LAYING THE TRACK

Trim the extra pipe from the ends and drill $\frac{3}{32}$" pilot holes opposite the cut at 1' to 2' intervals to hold the pipe firmly to the floor at the necessary radius. Make sure that the slot in the pipe runs along the top as the pipe is laid. Use two small shims to open the slot in the pipe, and screw the pipe to the deck with #8 pan head screws. These provide the greatest strength with the lowest profile. Washers may be added if desired, but make sure that they do not cause the slot to remain open after the shims are removed.

DOGS

The most effective dog is a simple knife of $\frac{1}{8}$" x 1" mild steel mounted vertically. Mount dogs along the movable unit as needed, allowing $\frac{1}{2}$" to ride inside the pipe. The polyethylene pipe will close on the dog and hold the unit firmly in place, limiting side-to-side movement. In situations where the moving unit will be subject to high speeds or rough treatment that could cause it to jump the track, bend the end of the dog at a 90° angle as in Figure 1.

This design was used to guide a 25-pound bench a distance of 40' along a track with a 12' radius. The show ran for forty performances without a mishap.

OTHER NOTES

This article describes the use of polyethylene pipe as a floor-mounted track, but the pipe is equally appropriate for overhead or vertical use in carefully planned and monitored applications.

It should be noted that the track has a relatively high profile (1") which limits its use to situations where it can either be masked or incorporated into the design.

Polyethylene pipe comes in coils and is difficult to manipulate when it is cold. Gently warming the pipe, *e.g.*, by placing it in the sun for a short time, will make it pliable and easy to work.

Polyethylene pipe is black and will not take paint without special chemical treatment. It is also resistant to most adhesives but may be heat-bonded.

Polyethylene is flammable, so check local fire regulations and consult the fire department before use.

❧❧❧❧

To fully conceal a trapdoor in a deck, the hinging mechanism must be hidden. However skillfully installed, the knuckle of a conventional hinge will protrude above deck level. A hinge developed to solve this problem can be mounted below the deck and works much like the hinge on an automobile hood. See Figure 1.

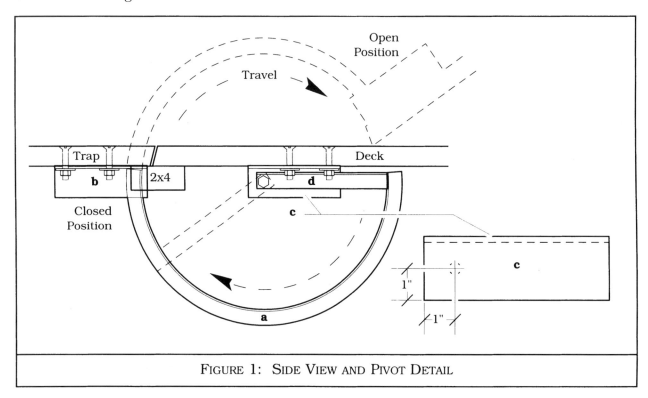

FIGURE 1: SIDE VIEW AND PIVOT DETAIL

HINGE CONSTRUCTION

Dimensions and materials listed were used for a trapdoor in a deck with 9" of clearance below. They are intended only as a guide and may be adjusted to suit other requirements.

MATERIALS NEEDED PER HINGE (LOWERCASE LETTERS REFER TO FIGURE 1)

a. 1" x 1" x $1\frac{1}{8}$" steel angle cut to 2'-1"
b. 2" x 2" x $\frac{3}{16}$" steel angle cut to 6"
c. 2" x 2" x $\frac{3}{16}$" steel angle (pivot) cut to 6"
d. 1" x 1" x $\frac{1}{8}$" steel angle cut to $8\frac{1}{2}$"

ASSEMBLY PROCEDURES

1. Bend **a** to an 8"-radius arc as shown in Figure 1. If appropriate bending tools are not available, straight pieces may be welded together provided that the arc they describe is big enough to clear the edge of the platform. If the arcs of paired hinges are not identical, whether bent or welded, the trapdoor will not open evenly.

2. Weld one end of **a** at a right angle to **b** as shown.

3. Weld the other end of **a** to piece **d** so that the pivot hole intersects the center of the arc.

4.   Bolt **c** to **d** at the pivot point, placing a washer between the pieces to reduce binding. Double-nut the bolt to prevent loosening.

INSTALLATION PROCEDURES (REFER TO FIGURE 2)

Note: For accurate placement, the trapdoor will be cut out of the platform after the hinges have been attached.

1.   Align the hinges parallel to each other in place beneath the deck.

2.   Bolt the hinges in place with countersunk stove bolts.

3.   Cut the trapdoor with a saw set at a 15° bevel so that the trapdoor will not bind on the platform.

4.   Frame the trapdoor opening beneath the deck to provide support.

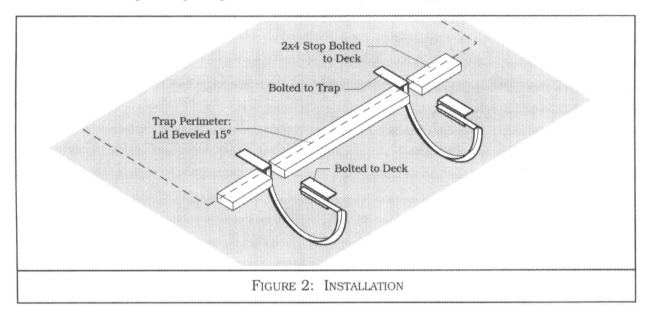

FIGURE 2: INSTALLATION

OTHER NOTES

1.   Moving the pivot point closer to the trapdoor will allow the trapdoor to open farther, but the open trapdoor will never lie flat on the stage floor.

2.   The hinge will work with irregularly shaped trapdoors.

3.   The hinge can also be used to hang concealed doors in vertical panels.

Nylon Rollers

Ray Forton

Using shop-built nylon rollers like those shown in Figures 1 and 2, a low-profile furniture or prop pallet can be constructed to an overall height of only $1\frac{5}{8}$". The pallet itself — an unframed layer of $\frac{5}{8}$" plywood skinned with $\frac{1}{4}$" Masonite® — makes up more than half of that height. The rollers add only $\frac{3}{4}$". Supporting a typical 4x8 pallet takes 12 to 15 rollers.

The roller is made of 1" nylon rod with a $\frac{1}{4}$" hole for the axle drilled through the center. The nylon rod is available at commercial plastics suppliers and costs about \$3.00 to \$4.00 per foot. $\frac{1}{4}$"-diameter cold rolled steel rod acts as the axle, and two pieces of $\frac{1}{8}$" x 1" x 3" flat steel are the mounting plates. Two $\frac{3}{16}$" washers will fit snugly around the $\frac{1}{4}$" steel rod and act as spacers between the roller and the mounting plates.

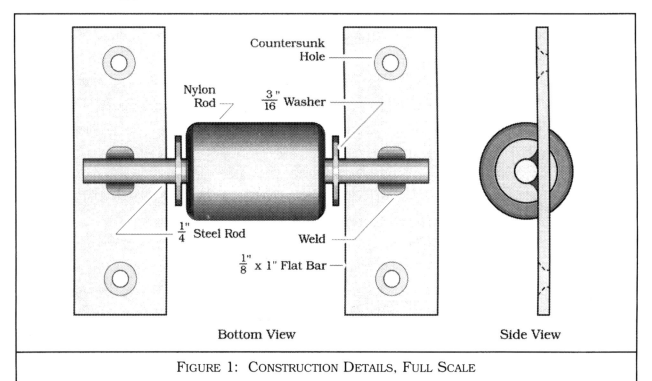

Countersunk Hole

Nylon Rod

$\frac{3}{16}$" Washer

$\frac{1}{4}$" Steel Rod

Weld

$\frac{1}{8}$" x 1" Flat Bar

Bottom View

Side View

FIGURE 1: CONSTRUCTION DETAILS, FULL SCALE

CONSTRUCTION

1. Using a bandsaw with a fine-toothed blade, cut the nylon rod into $1\frac{1}{2}$" lengths; sand the ends of each roller, slightly rounding off the ends for smoother rolling.

2. Drill a $\frac{1}{4}$" hole through the center of the rod to accept the axle. It is important to center the hole exactly so that the roller will spin evenly. Accurate centering can be achieved by mounting a drill chuck in the tail stock of a lathe, clamping the rod in the lathe's head stock, and feeding a twist drill into the end of the rod. With an appropriate clamping jig, a drill press can also be used with good results.

3. Drill a hole in each end of the pieces of flat bar; countersink the holes to accept flat head wood screws.

4. Assemble the rollers as shown in Figure 1, placing the washers on the axles between the nylon rod and the flat bar. Allow just enough space between the plates for the roller to spin freely.

5. Spot weld the flat bar to each end of the axles. Be careful: too much heat will melt the nylon.

6. To prepare the plywood to receive the assembled rollers, cut holes which will provide clearance around the rollers. Screw the plates onto the plywood with flat head wood screws.

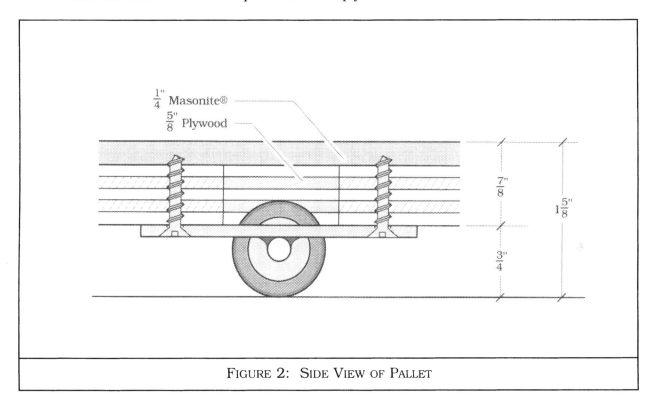

FIGURE 2: SIDE VIEW OF PALLET

NOTE

Given their small diameter, the rollers must spin very fast when moving and, thus, are slightly louder than normal casters. The resulting noise is not usually a problem, but its consequences must be considered.

❧❧❧❧❧

Nylon Rollers Modified

Thomas G. Bliese

The preceding *Technical Brief* design for nylon rollers shows the axle welded in place on the bottom of the supporting flat bar. If the axle were welded instead to the top of the flat bar as shown in Figure 1 here, the pallet could be made thinner by $\frac{3}{8}$".

Those who attempt this second design will need to observe the following points:

1. The loss of $\frac{3}{8}$" between the pallet and the bed on which it rides requires that the bed be cleaned thoroughly and frequently. Further, deflection of the pallet may require the use of more rollers in this design than in the original.

2. Adequate clearance between the roller and the Masonite® skin must be provided by routing a slight depression into the bottom of the Masonite®.

3. The quality of the weld between axle and flat bar is more critical in this design, which adds a tensile load to the original design's shear load. Since the nylon rod will melt under too long an exposure to high temperatures, speed and accuracy in welding are essential.

Both the original and modified designs appear below.

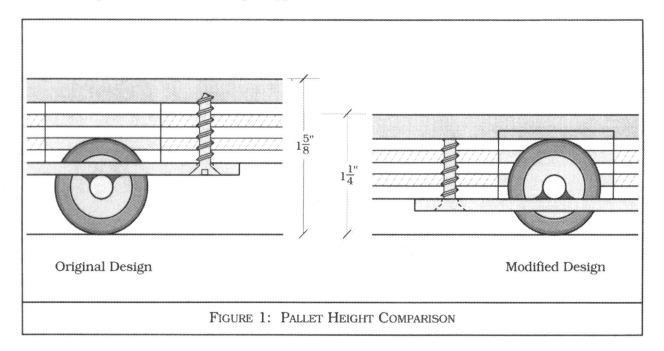

Original Design

$1\frac{5}{8}$"

$1\frac{1}{4}$"

Modified Design

FIGURE 1: PALLET HEIGHT COMPARISON

Most commonly, technicians see fixed and swivel casters as the only alternatives for rolling a platform across the stage. Even the more versatile of these, the swivel caster, has limitations when a change of direction is required. In reversing directions, the swivel caster tends to pivot around its point of contact with the floor, causing a familiar and unwelcome lurch.

The problem is avoidable, however. The solution lies in the use of the zero-throw caster, which allows a platform to move easily in any direction without the "throw" of the standard swivel caster. See Figures 1A through 1C.

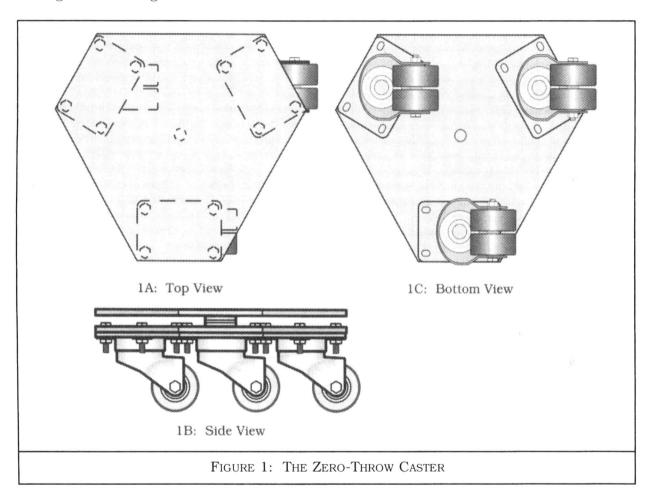

1A: Top View   1C: Bottom View

1B: Side View

FIGURE 1: THE ZERO-THROW CASTER

DESCRIPTION AND FABRICATION

In a zero-throw caster, three casters are attached to the bottom of a steel caster plate. The caster plate is then attached to the platform by way of a thrust bearing and a steel mounting plate which allows the caster plate to swivel independently of the platform.

Cut a $\frac{1}{4}$" steel caster plate big enough to permit the casters to swivel freely. Though Figure 1 shows the caster plate as an equilateral triangle with the vertices cut off, the shape is optional. Drill $\frac{5}{16}$"-diameter holes in the mounting plate for attachment to the platform. Make certain they are near enough the outside of the mounting plate to permit tightening the bolts after the whole caster unit is together. Drill 12 holes in the caster plate for mounting the casters, as well as a hole in the center to receive a shoulder washer through which a $\frac{1}{2}$" steel pivot rod will pass. Attach the casters to

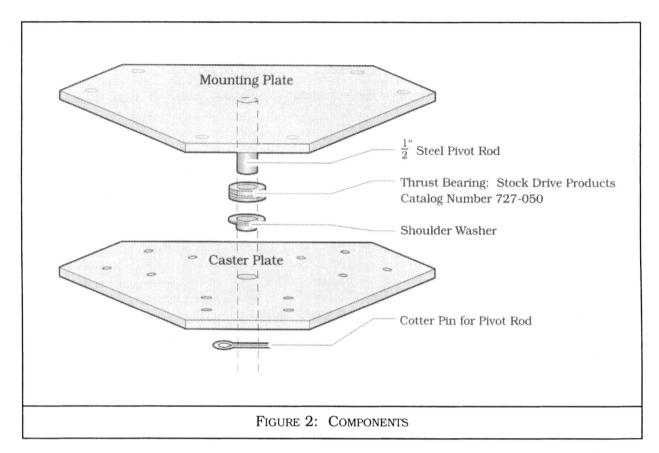

FIGURE 2: COMPONENTS

the caster plate and insert the shoulder washer in the caster plate's central hole. Cut a $\frac{1}{2}$" steel pivot rod to a length of 2" and drill a hole through one end to receive a cotter pin that will be $\frac{1}{8}$" below the caster plate after assembly. Weld the other end of the pivot rod to the center of the mounting plate. Slide first the thrust bearing and then the caster plate (with the shoulder washer) over the pivot rod, and secure all three pieces together with a cotter pin. The components and assembly are shown in Figure 2.

NOTES

1. This zero-throw caster can carry three times as much weight as any one of its swivel casters and is, therefore, quite useful in high-load applications.

2. The larger a zero-throw's footprint is, the more easily it will turn. On the other hand, the larger it is, the heavier and more awkward it is to handle and store.

3. Platforms that ride on zero-throw casters cannot be framed in the familiar 2x4-perimeter fashion. Platform framing designs must transfer loads to the zero-throw casters' central pivots and must accommodate their mounting plates.

❧❧❧❧❧

Many sets require concealed or semi-concealed hinge joints for trapdoors or secret wall panels. But scene shops often lack the time or resources to provide the elaborate and sometimes costly hardware often used to accomplish such joints. The alternative described here is an easily made, inexpensive hinge which can fold flat against itself and yet be entirely concealed.

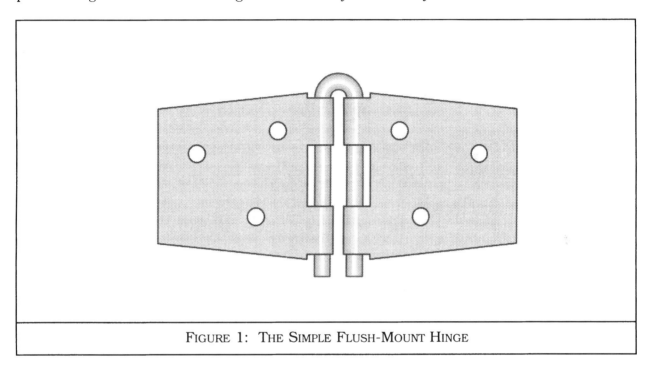

FIGURE 1: THE SIMPLE FLUSH-MOUNT HINGE

As Figure 1 illustrates, the leaves of this hinge are the female leaves taken from a pair of identical stock hinges. An appropriate diameter steel rod bent into a "U" forms a double pin connecting the leaves. This pin allows the leaves to fold flat against each other when the trapdoor is open, and keeps the hinge barrels from protruding above the surface when it is closed. The two positions are shown in Figures 2 and 3.

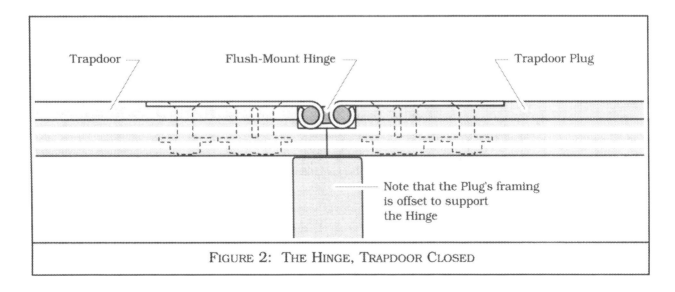

FIGURE 2: THE HINGE, TRAPDOOR CLOSED

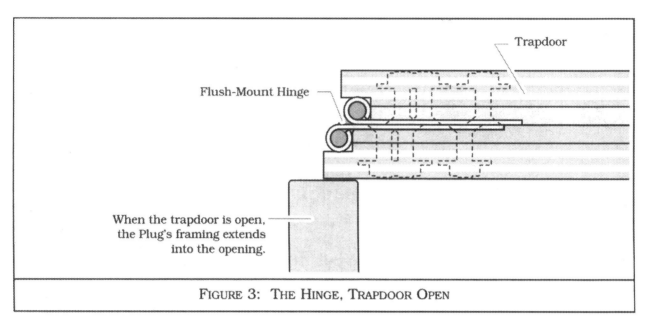

Flush-Mount Hinge

Trapdoor

When the trapdoor is open, the Plug's framing extends into the opening.

FIGURE 3: THE HINGE, TRAPDOOR OPEN

NOTES

1. Since the hinge is installed face down, *i.e.*, with the barrels extending into the work, a groove for their barrels must be routed into the work.

2. In many installations, the groove itself may serve to keep the pin from sliding out of the barrels; but to hold the pin securely in the hinge, bend, weld, or cotter pin the protruding ends to close the loop.

3. To make the hinge completely flush, it is mortised into the work and the bolt holes are counterbored to accommodate nuts and washers.

❧❧❧❧

At times, the simple solution is the most enjoyable. I developed a number of ideas for the construction of a hinge for a two-panel elevator door, all of which relied upon the principles of geometry and many of which were quite elaborate. We adopted the simplest of these and were quite pleased with the results. In our production both panels of an elevator door retracted simultaneously into a pocket in the wall stage left of the door. The panels, which covered a 4'-wide opening, were built as hard-covered 1"-tube-steel frames, designed to overlap each other by 3" at the center. In order to complete their travel simultaneously, the SR panel had to move twice the distance covered by the SL panel. A pantograph hinge allowed us to accomplish this movement.

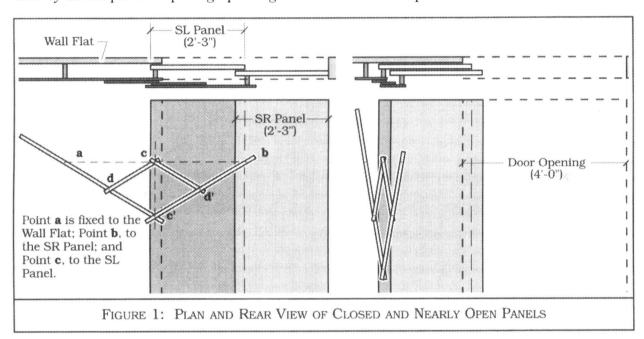

FIGURE 1: PLAN AND REAR VIEW OF CLOSED AND NEARLY OPEN PANELS

For a hinge like this to work, pivot points **a**, **b**, and **c** (Figure 1) must lie on a line parallel to the panels' line of travel. Working on an elevation showing the panels completely open, draw line **ab** between the farthest two pivot points and erect its perpendicular bisector. The intersection of those two lines indicates the location of pivot point **c**. Next, from pivot point **a**, draw a line downward and across the perpendicular bisector. This line must be longer than the distance between **a** and **c** or the panels won't close completely — and it must cross the bisector above the floor line or they won't open completely. The intersection of this line and the perpendicular bisector of **ab** indicates the location of pivot point **c'**. Pivot point **d** will be found halfway between **a** and **c'**; pivot point **d'**, halfway between **b** and **c'**.

In our application, arms made of 1" x $\frac{1}{8}$" steel bar stock flexed easily enough to permit snug bolting at the pivot points, and once they had been joined, they were sturdy enough to move the panels without bending. The addition of a counterweight to the offstage arm kept the panels closed.

DESIGNED WITH

Alan Hendrickson.

Handle-action toggle clamps provide a sure and inexpensive means of securing trapdoors in plat-forms and stage floors. Such clamps are basically locking levers, which require very little effort to open or close and yet offer substantial holding force when they have locked into their closed posi-tion.

By attachment to a low-pressure double-acting pneumatic cylinder with a 5" to 8" piston throw, an inexpensive, manually operative toggle clamp can be converted into a remote-controlled trapdoor lock. The toggle clamp is easily modified by drilling a hole through the handle and pinning the pis-ton to the handle using a female rod-clevis assembly available from pneumatics suppliers. In-shop modification for pneumatic operation costs far less than the $80.00 that commercially produced pneumatic toggle clamps cost. It also offers the additional advantages of manual override in case the pneumatic system fails and reusability of component parts after disassembly. Figures 1 through 3 illustrate the clamps and show their use as toggle-clamp locks.

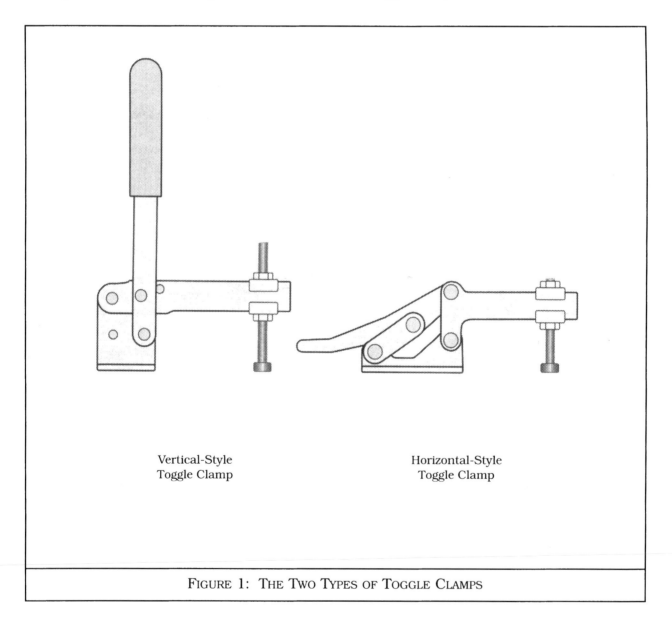

Vertical-Style
Toggle Clamp

Horizontal-Style
Toggle Clamp

FIGURE 1: THE TWO TYPES OF TOGGLE CLAMPS

MODIFICATION AND OPERATION

Vertical toggle clamps are generally simpler to modify but require considerably more lateral operating space than their horizontal style counterparts. When horizontal toggle clamps are to be used, care must be taken in the placement of the holes in the handle. The travel of the horizontal clamp's handle describes an S-shaped path, and if the hole is not properly located, the cylinder will not open and close the clamp.

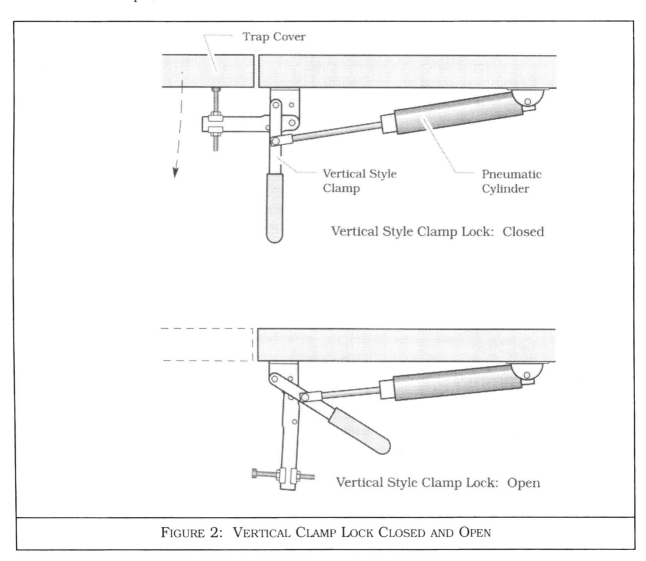

FIGURE 2: VERTICAL CLAMP LOCK CLOSED AND OPEN

Operation of the clamp-lock system is simple, as one four-way switch will control a lock or several locks plumbed in parallel. Visible and audible indicators of closure are available. Consideration should be made in advance for the noise the pneumatics will make. There are two primary causes of noise in the system: the clamp hitting the trapdoor during engagement, and the exhaust of the cylinder when releasing the lock. The construction of the trapdoor will determine how loud the sound of engagement will be. Cylinder exhaust noise may be overcome by installing a muffler on the exhaust line and running the exhaust line as far from the audience as possible.

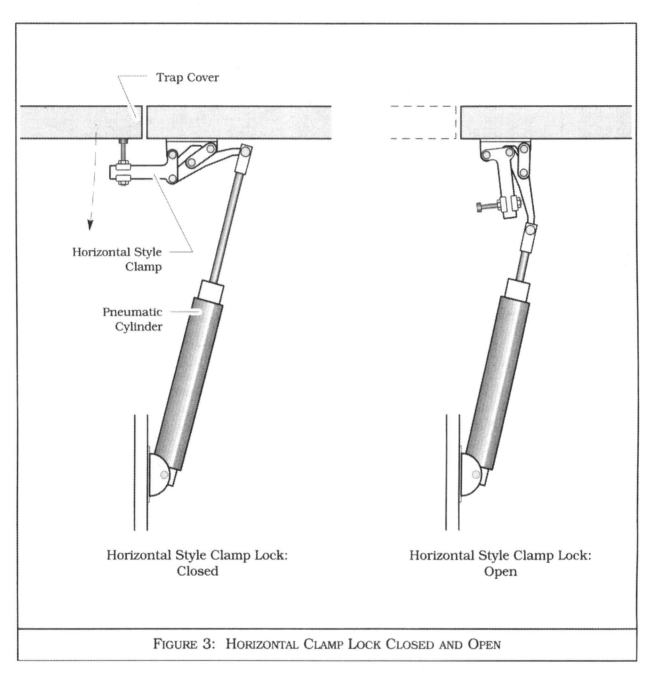

Trap Cover

Horizontal Style
Clamp

Pneumatic
Cylinder

Horizontal Style Clamp Lock:
Closed

Horizontal Style Clamp Lock:
Open

FIGURE 3: HORIZONTAL CLAMP LOCK CLOSED AND OPEN

Toggle clamps can be purchased from McMaster-Carr, and pneumatic cylinders and supplies from Clippard-Minimatic, Inc.

For a production of *Moon Over Miami* at the Yale Repertory Theatre, we devised a special stage jack to support an 8' flat sitting on the back of a 4' x 8' oval wagon. See Figure 1. As seen from the side, the unit was an unsupported "L" and had to be secured so that the flat would not move when actors walked on the wagon. This meant that the jack had to lock into place in a plane that was perpendicular to the plane of the flat. In addition, limited backstage space made it essential that the jack be either removable or able to fold away so that the unit could store up against a wall.

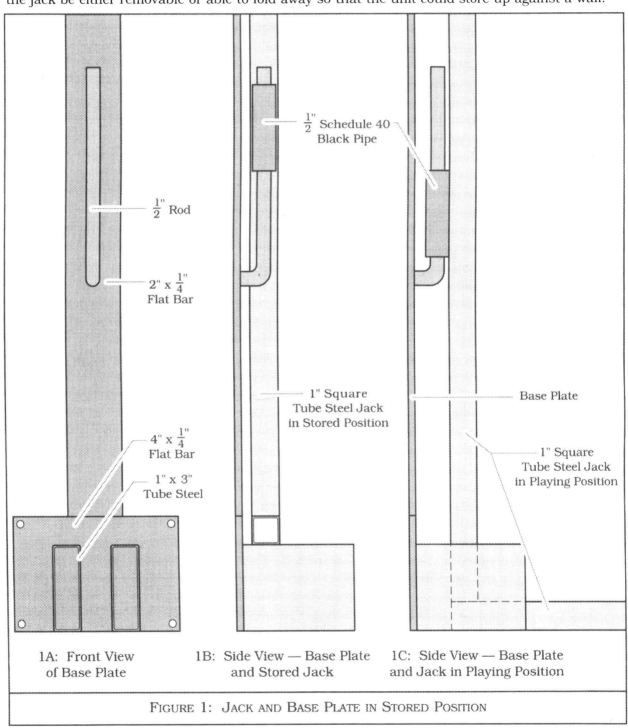

$\frac{1}{2}"$ Schedule 40 Black Pipe

$\frac{1}{2}"$ Rod

$2" \times \frac{1}{4}"$ Flat Bar

1" Square Tube Steel Jack in Stored Position

Base Plate

1" Square Tube Steel Jack in Playing Position

$4" \times \frac{1}{4}"$ Flat Bar

1" x 3" Tube Steel

1A: Front View of Base Plate

1B: Side View — Base Plate and Stored Jack

1C: Side View — Base Plate and Jack in Playing Position

FIGURE 1: JACK AND BASE PLATE IN STORED POSITION

A Quick-Locking Jack

C. Ken Cole

Jacks and stage braces have been around a long time. Our problem was to make one that locked into playing position quickly — and could be struck just as quickly. Several different options were considered, and we finally chose the pivoting steel-frame design, because of its simplicity. The jack is a two-part system: the jack itself and its base plate.

THE JACK

Except for two 3"-sections of $\frac{1}{2}$" pipe welded to its upright, the jack is a typical assembly of 1" square tube steel. A $\frac{1}{2}$" hole through the foot of the jack accepts a threaded pin that secures the jack to the stage floor. The pieces of $\frac{1}{2}$" pipe act as hinge barrels that let the jack swing back and forth from storage to playing position.

THE BASE PLATE

The base plate is constructed of steel rod, flat bar, and 1" x 3" tube steel. See Figure 1. An upright of 2" x $\frac{1}{4}$" steel plate and two short sections of 1" x 3" tube steel separated by a $1\frac{1}{8}$" gap are welded to a short length of 4" x 6" x $\frac{1}{4}$" flat bar. The pieces of tube steel lock the jack into playing position. The long upright of 2" x $\frac{1}{4}$" flat bar above these pieces carries two bent pieces of $\frac{1}{2}$" steel rod that act as hinge pins for the jack. Located about 1' from the top and bottom of the base plate, the bent-rod pins must extend far enough from the base plate to allow the jack to swing freely.

ASSEMBLY AND OPERATION NOTES

After the completed base plate has been bolted to a unit, the jack's hinge barrels are lowered onto the bent-rod pins. When the jack swings perpendicular to the flat, it slips into position between the two pieces of 1" x 3" tube steel and locks into place. Lifting up on the jack unlocks it, allowing it to swing to either side for storage. Greasing the hinge parts helps assure smooth and quiet movement. In its storage position, the jack can be held in place with a simple hook and eye latch.

This design can be modified very easily to fit a variety of flats. Careful construction will result in tolerances tight enough to move smoothly without rattling.

❧❧❧

Scenery Materials

Corrugated (Kraft) Cardboard as a Scenic Material   *John Robert Hood*

Corrugated cardboard has been used occasionally as a scenic covering and profile material. As commercially available, this material is a marginal substitute for conventional scenic materials. A critical disadvantage in scenic use is the high degree of flammability: because of the nature of the materials and construction, conventional flame-retarding treatments are not successful when applied to corrugated cardboard. This material should not be used where fire and safety code limitations require flame-resistant characteristics, nor should it be used where burning materials would create a fire safety hazard.

"Three-ply corrugated craft" is the form that has been most successfully applied. This material may be procured from manufacturers of corrugated shipping boxes or distributors of packaging materials; occasionally, "second" or "reject" sheets are available at little or no cost. Sheet sizes vary from manufacturer to manufacturer, but generally fall within the 4' x 8' to 5' x 10' range. Waterproof (oil- or wax-treated) forms should be avoided, as painting is difficult and flammability is increased.

This material may be cut with conventional scenic tools, such as matte knives, Cutawls®, and table saws. Cutting blades should be of a fine-toothed or knife form to avoid tearing and "chewing" the fragile paper surfaces. The bonding glues used in the manufacture of cardboard are usually abrasive and tend to dull cutting tools quickly. The material is easily damaged by crushing — exercise care in handling.

Cutting and profiling should be completed before the cardboard is attached to a frame. The side with the finer corrugations should be used as the finished surface. Use large-headed nails to secure the material to framing members; avoid "dimpling" the material by driving the nail too deep. Carpenter's glue or white glue may be used to attach the cardboard to the framing. Even pressure must be applied to hold the joint while the glue sets — without pressure, the cardboard will tend to warp away from the frame. Scenic elements can be fabricated without wooden framing by folding and gluing the cardboard into box-like structures; reinforcements and stiffeners can be formed by laminating strips of cardboard. Visible seams can be dutchmanned with Kraft® paper tape or with gauze and wheat paste. Edges should be covered to prevent tearing and to conceal the ragged corrugations.

Painting techniques must be adjusted to the material, as it warps and bows readily when surfaces are moistened. Also, the water-soluble glues used in manufacture soften, and the corrugations delaminate when wetted. Generally, "puddling" should be avoided; paint should be mixed thin and applied quickly. If over-painting is necessary, allow sufficient drying time so that moisture does not penetrate the glue bond. Framing and reinforcement should be placed to avoid large unsupported areas. The material should be back-painted first, then quickly turned and front-painted: the simultaneous wetting and drying of both surfaces creates opposing bowing and warping stresses, providing the moisture does not penetrate and loosen the glue bonds.

While corrugated cardboard offers advantages of cost, weight, and time, its disadvantages require more care in handling and painting, and shop techniques must be adjusted to achieve satisfactory results. This material should not be used if flame-retardancy or long-term durability is required.

৶Ꭿ৶Ꭿ৶Ꭿ

Having become used to working with a variety of standard mastics and adhesives, theatre technicians are prone to overlook double-stick foam tape as a useful construction material. Such tape can create a quick, tight seal between surfaces and can be used almost anywhere that more traditional bonding materials may be difficult or impossible to use.

APPLICATIONS

Double-stick foam tape can eliminate many of the squeaks in wood-framed, plywood-covered platforms. Before attaching the plywood, simply apply double-stick tape to the top edges of the frame so that it acts as a gasket, preventing the cover from moving against the frame. In building steel-framed, Plexiglas®-covered platforms, use of the tape eliminates the cracks and crazing that result from the use of screws. In this application, the tape has the added advantage of allowing the Plexiglas® to expand and contract as necessary.

Double-stick tape is useful in other Plexiglas® applications as well. For example, laying the tape along the back edge of a window frame and simply pressing a pane of Plexiglas® in place finishes a window very quickly. Similarly, the foam tape can be used to apply wood mouldings to Plexiglas® surfaces. The quality of the bond between Plexiglas® and other materials can be easily checked by inspecting the appearance of the applied tape: wherever it appears compressed, the bond is sound.

Finally, double-stick foam tape has few equals in attaching Ethafoam® rod and Styrofoam® to scenery and is especially useful in attaching anything that has to conform to a curve, since it immediately achieves flexible adhesion without the use of clamps.

One note of caution: foam tape will hold approximately 1psi in tension. Thus, though it is useful in resisting either compression or shear forces, it should not be used to attach objects to ceilings or in other tensile applications.

HANDLING

1. The presence of dirt poses the greatest difficulty in the use of this product. All surfaces to be taped must be clean — wood surfaces must be free of dust; metal surfaces, free of greasy dirt.

2. As a rule, the rougher the surface to be taped, the thicker the tape must be. Tape $\frac{1}{16}$" thick is recommended for use with wood.

3. Normally, applying small pieces of tape at evenly spaced intervals provides the required holding force. When using a continuous strip to satisfy visual demands or to form gaskets, however, roll out a three-foot section and press it firmly into place before moving on.

4. The best "strike" approach to separating pieces that have been taped together is to split the foam with a razor blade. The remaining tape can be scraped or rubbed off any hard surfaces.

PRODUCT INFORMATION

Double-stick foam tape is made by several manufacturers, 3M being the largest. It is commercially available in thicknesses of $\frac{1}{32}$", $\frac{1}{16}$", and $\frac{1}{8}$". Though commonly available in widths ranging from $\frac{1}{2}$" to 2", it is manufactured in widths up to 54".

Foam tapes for general use are made from polyethylene foam and coated with either rubber or acrylic adhesive. The rubber adhesive, typically found on tapes available from art and office supply stores, picture framing shops, and department stores, sets up quickly and cannot be pulled up and relaid once it has been pressed into place. The acrylic adhesive, which can be obtained in large quantities from bulk paper and industrial tape suppliers, takes longer to reach full bond strength and can be taken up and relaid for up to twenty-four hours after application. Both adhesives stick to wood, plastic, metal, and plastered or painted surfaces.

❦❦❦

A cross section like that shown in Figure 1 poses no problems for the set builder — as long as it represents a straight piece. But the staircase and handrail that appeared in a Yale Repertory Theatre production of *A Play of Giants* were not straight: the 5'-wide unit ran through two different curves over its 8' rise. Thus, we had to find either a way to bend strong but normally straight materials into a compound curve or some sort of casting material that would develop the necessary strength.

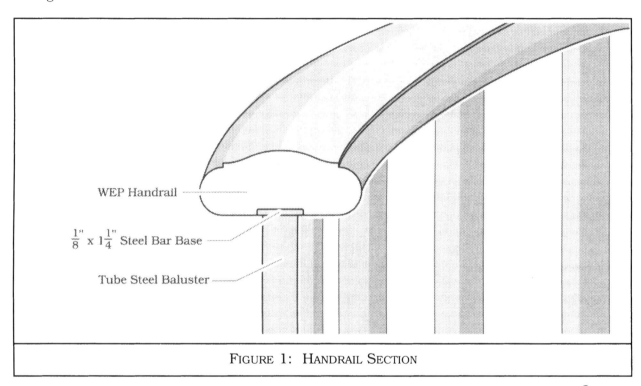

WEP Handrail

$\frac{1}{8}$" x $1\frac{1}{4}$" Steel Bar Base

Tube Steel Baluster

FIGURE 1: HANDRAIL SECTION

Discarding the former approach as too time-intensive, we turned to the use of "Water Extended Polyester" (WEP #662P). WEP is sold as an emulsion-catalyst casting mix. Once cured, its tensile strength is comparable to that of #2 pine at around 1450psi; it machines well; it accepts nails and screws with proper pilot holes; it sands easily to a smooth surface; and it takes most paints. But its greatest advantage lies in the fact that, as WEP cures, there is a short period within which a still-pliable casting can be pulled from its mold and formed into curves while retaining its cross-sectional shape. By following the process described below, we were able to cast the handrail in an easily built straight mold and then bend the straight casting into the required curve.

THE MOLDS

We built two molds — a 2'-long control mold and a 16'-long production mold — having the cross section shown in Figure 2. We used the control mold in determining the proper emulsion-to-catalyst ratio and batch timing. We made the production mold as long as practicable and yet short enough to keep the casting from becoming unwieldy during bending.

Aware of WEP's machinability, we decided to round the bottom edges of the handrail by sanding the cured casting rather than to complicate the molding process. The form for the gentle swell on the top of the handrail was made of plasticene shaped with a template and smoothed with a heat gun. A hand-rubbed coat of paste wax served as a mold release.

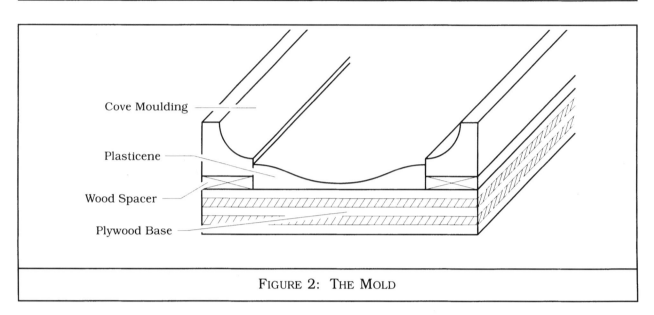

Cove Moulding

Plasticene

Wood Spacer

Plywood Base

FIGURE 2: THE MOLD

TESTING, CASTING, AND FOLLOW-UP

As the mixture set up, the casting temperature approached 200°F, necessitating gloves for handling. Five to six minutes after we had poured the evenly mixed WEP into the control mold, we were able to press a flat centerline trough into the bottom of the handrail casting. The trough helped locate the handrail on its structural support — a curved length of steel bar welded to the tops of the tube-steel balusters, which were already in place on the staircase unit.

At carefully timed intervals after each successive test pour, we pulled test castings out of the mold. Those pulled too soon broke because the resin had not yet developed enough tensile strength. Those left in the control mold too long proved too stiff to bend. By keeping careful records of each control-casting, we were able to discover the right molding time and batch proportions.

Once we had established a successful process, we began production casting. As each of these longer sections set up, we pulled it from the mold, supporting it at 2' to 3' intervals, and carried it to the balustrade as a straight piece. Then, having positioned one end of the casting's trough on its steel-bar base, we easily bent the casting into its final shape. We found that, after another thirty minutes, we could remove the few clamps we had used to hold the handrail in place, and that, after waiting twenty-four hours, we could machine the casting.

PROPERTIES AND COST

Instructions for mixing and using WEP appear on the side of the five-gallon pails in which it is sold. When ordering WEP #662P from its manufacturer, Ashland Chemicals, ask for all the available product information — not just the mixing instructions. A number of important variables involved in such unorthodox WEP uses as this can become clear only with hands-on experience. The 1% shrinkage that occurs during the curing process, for instance, could be a critical factor in deciding whether WEP is an appropriate choice for a different theatrical application.

Unlike other polyester resins that are used full strength, WEP's volume is extended by the addition of water. Adding water, of course, decreases the cost: a five-gallon pail of WEP and a pint of catalyst makes approximately seventy feet of the handrail shown, and costs about $70.00.

WARNING NOTES

The ingredients in WEP — as in any polyester resin — are extremely dangerous and must be handled intelligently. Familiarize yourself with the potential hazards and the necessary precautions. The resin itself contains styrene, which produces explosive vapors at room temperature. Use positive general ventilation and, if you stir with an electric drill, use a paint mixer blade with a long shaft to avoid setting off an explosion. The catalyst contains methyl-ethyl ketone peroxide, which contains enough free oxygen to become explosive if heated or left in bright sunlight. If the catalyst were accidentally mixed with acetone, a common polyester solvent, it would form an even more explosive mixture. Keep acetone out of the WEP work area, where it could be used by mistake. Never use acetone for cleaning MEK peroxide or catalyzed emulsion from any surface.

These chemicals also present serious health hazards ranging from blindness on contact to liver damage. Never reuse empty WEP containers for any purpose: process them as Hazardous Waste.

Despite its potential for harm, WEP can be used safely by those who wear splashproof goggles, gloves, and protective skin cream, and who frequently change the organic vapor cartridges in their respirators.

For additional information, consult the following sources:

Ashland Chemical Company   Industrial Arts Supply Co.
Polyester Division   1408 West Lake Street
P.O. Box 2219   Minneapolis, MN 55408
Columbus, OH 43216   ATTN: Michael J. Raymond
(manufacturer)   (distributor)

Reichold Chemicals   *Artist Beware*
525 North Broadway   Dr. Michael McCann, author
White Plains, NY 10603   Watson-Guptill Publications
(MEK peroxide)   (health hazards)

❧❧❧❧

Comparing Four Plastics as Scenery Glides

Edmund B. Fisher

Plastic glides are often used instead of casters as aides in shifting scenery, especially when casters would be aesthetically objectionable. But not all plastics work equally well as glides.

This article presents a ranking of four plastics — Nylon, Delrin®, Nylatron GS®, and Virgin Teflon® — as glide materials. The four were chosen for a variety of reasons: Nylon, because it is often used as glide material; Teflon®, because it is often assumed to be more effective than Nylon; and Delrin® and Nylatron GS®, because local distributors suggested them as possible substitutes. The four materials are compared in terms of their cost, workability, and coefficients of friction.

The first two of these points of comparison are almost self-explanatory: cost rankings were determined by averaging price quotes from local suppliers; and workability rankings, by observing how easily the materials could be cut and drilled. The third point of comparison, coefficients of friction, is discussed below.

COEFFICIENTS OF FRICTION

Coefficients of friction are indicators of relative slipperiness. The formula for static friction is $F_S = \mu N$, where

> F_S = the amount of force required to overcome static friction, *i.e.*, to cause movement to start,
> N (Normal force) = the weight of the object being tested, and
> μ = the coefficient of static friction.

In general, the lower its coefficient of static friction, the more easily one material will begin to slide across another. Further information about coefficients of friction can be found in *Machinery's Handbook* and other standard references.

THE TEST

Figure 1 illustrates the test setup. For the test, 1"-diameter rods of the plastics to be tested were bandsawn into half-rounds and cut into 3" lengths. Two countersunk holes drilled through each glide permitted its attachment to a plywood test bed by means of drywall screws. Figure 2 illustrates a finished glide.

The test bed was placed glide-down on a level test surface, and weight W_1 was placed on the test bed. Next, the test bed was joined to hanging weight W_2 by a length of manila rope. In order to keep the rope's strands from catching on the edge of the table and interfering with the results, the rope was passed over a pipe fastened to the edge of the table.

Testing consisted of gradually increasing weight W_2 until W_1 and the test bed began to slide across the test surface. Movement began when the weight of W_2 was equal to F_s, the force needed to overcome static friction.

Finally, a coefficient of static friction for each combination of glide material and test surface was determined by rearranging the static friction formula and substituting the empirically derived values for W_2 and W_1. Thus, the coefficients were derived using the formula $\mu = W_2/W_1$.

Each set of glides was tested on three different test surfaces: lauan, painted Masonite®, and dirt-covered Masonite®.

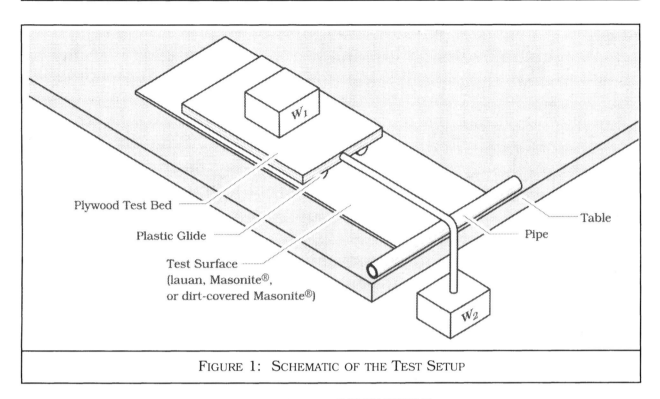

FIGURE 1: SCHEMATIC OF THE TEST SETUP

Plywood Test Bed

Plastic Glide

Test Surface
(lauan, Masonite®,
or dirt-covered Masonite®)

Table

Pipe

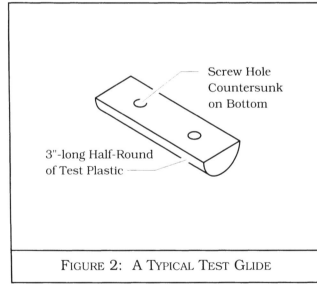

Screw Hole
Countersunk
on Bottom

3"-long Half-Round
of Test Plastic

FIGURE 2: A TYPICAL TEST GLIDE

CONCLUSIONS

Table 1 presents the derived coefficient of static friction and comparative ranking for each glide material. Table 2 summarizes the friction, cost, and workability rankings and averages them in an overall ranking.

The results of the tests on Nylatron GS® were disappointing, as the table reveals. Surprisingly, Teflon® came in a distant third. It readily picked up damaging particles (especially metal) in the shop, and therefore performed poorly in the dirt test.   Nylon performed second-best overall, though its performance varied significantly from one test surface to another. In terms of cost, workability, and coefficient of friction, Delrin® was the clear winner.

Material	Lauan	Dirt	Masonite®	Average	Ranking
Virgin Teflon®	0.373	0.556	0.679	0.536	3
Delrin®	0.317	0.489	0.528	0.445	1
Nylon	0.339	0.500	0.755	0.531	2
Nylatron GS®	0.469	0.556	0.604	0.543	5
TABLE 1: STATIC-FRICTION-TEST RESULTS					

Material	Friction	Cost	Work-ability	Overall
Virgin Teflon®	3	4	2	3
Delrin®	1	2	1	1
Nylon	2	1	3	2
Nylatron GS®	4	3	4	5
TABLE 2: SUMMARY RANKINGS				

A FINAL WORD

Insufficient trials were conducted to evaluate these plastics' durability. Subjectively, however, Nylatron GS® seemed toughest, followed by Nylon and Delrin®. Teflon® seemed the most fragile.

These plastics seem to be the most commonly available choices, but are by no means an exhaustive sample of the thousands of possible glide materials. Individual users are encouraged to perform their own tests.

TEST MATERIALS

      Nylon 101 (MilSpec L-P-410a).
      Nylatron GS® (MilSpec L-P-410a).
      Delrin® (FedSpec L-P-392A, Type 2, and/or ASTMspec D2133-64T, grade 2).
      Virgin Teflon® (MilSpec Mil-P 19468).

Mr. Edmund Fisher's *Technical Brief* article, "Comparing Four Plastics as Scenery Glides" inspired the technical staff at Wilkes College to try plastic glides. At our local plastics distributor's suggestion, we used ultra-high-molecular-weight polyethylene (UHMWPE) in place of Mr. Fisher's "clear winner," Delrin®. The UHMWPE glides worked so well for us that they warranted further scrutiny. This article reports the results of our investigations.

Mr. Fisher's test sought to compare the coefficients of static friction of four plastics on three fairly standard deck materials. We wanted to test the coefficients of kinetic friction of plastics we are likely to use. While the coefficient of static friction indicates how much effort is required to induce two materials to begin sliding past each other, the generally lower coefficient of kinetic friction indicates how much effort is required to sustain movement once it has begun. Our experience in pushing scenery suggests that a stagehand who can't necessarily start a wagon moving can often keep it moving once it is under way. Technically stated, it is harder to overcome static friction than kinetic friction.

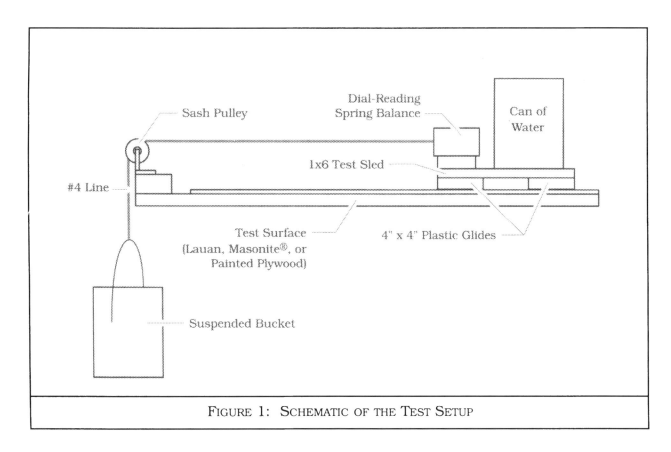

FIGURE 1: SCHEMATIC OF THE TEST SETUP

OUR TEST

Though interested in comparing UHMWPE to Mr. Fisher's "clear winner," we did not test Delrin®. Instead of E. I. Du Pont De Nemours and Company's acetal, Delrin®, our plastics supplier stocks the Polymer Corporation's acetal, Polypenco®. For practicality, we used Polypenco® instead of Delrin®. The plastics were tested on clean lauan, clean tempered Masonite®, and fir plywood painted with Para-Casein® Raw Umber and Burnt Umber.

Figure 1 shows our test rig, which differed from Mr. Fisher's in one respect. Mr. Fisher passed the rope joining his test sled and weight "over a pipe fastened at the edge of the test bench." Thus, his data includes not only the friction between his tested materials, but also the friction between manila rope and steel pipe. We used a sash pulley in place of the pipe to reduce extraneous friction, and we took our measurements at the sled.

Our test sled was a 12-inch piece of 1x6 weighted with a can of water and a spring balance. Two four-inch squares of the plastic being tested were screwed to the bottom of the 1x6. The screws were countersunk and the edges of the squares were chamfered. The test sled weighed 1720 grams for the Polypenco® acetal, and 1680 grams for the UHMWPE.

Our procedure was to put weight in the suspended bucket to tension the sash cord and gently tap the sled to get it to move without accelerating visibly. If the sled didn't keep moving, we added a little weight and tried again. We found that the sash-cord tension needed to keep the sled moving was remarkably consistent. Variation from trial to trial was less than the resolution of the spring balance.

The coefficient of kinetic friction was determined by dividing the weight of the test sled into the tension needed to keep it moving. The same spring balance was used in all the measurements, so that the absolute accuracy of the balance would not affect the results.

OUR TEST RESULTS

Table I presents our test results.

Material	Lauan	Masonite®	Painted Plywood
Polypenco® acetal	0.22	0.20	0.34
UHMWPE	0.22	0.17	0.33
TABLE I: KINETIC-FRICTION-TEST RESULTS			

CONCLUSIONS

There seems to be no significant difference between UHMWPE and Polypenco® in terms of coefficients of kinetic friction, and only minor differences in their other physical properties. Both plastics can be worked with standard tools in the scene shop. UHMWPE has a low melting point, and will melt if dull cutting tools are used, while dull tools seem to cause fewer problems with Polypenco®, which has a much higher melting point. At least locally, however, there is a significant difference in price. Pena-Plas of Scranton, PA, charges \$143.04 (\$4.47 per square foot) for a full 4' x 8' sheet of $\frac{1}{4}$" UHMWPE, and \$137.60 (\$17.20 per square foot) for a full 2' x 4' sheet of $\frac{1}{4}$" Polypenco®. Less than full-sheet quantities cost 50% more per square foot, but even at that higher price, UHMWPE is much less expensive than Polypenco® acetal. Thus, though both plastics are effective as scenery glides, UHMWPE is a bargain.

❧❧❧

Structural fiberglass consists of glass fibers and polyester resins combined in a variety of cross-sectional shapes once available only in steel or aluminum. The available shapes include round and square tube, I-beam, angle, channel, rod, bar, and flat sheet. For most theatrical needs, the standard series 500 is sufficient. Although this series is not inherently fire-retardant, it can be treated in the same manner as lumber to achieve flame retardancy. If flame retardancy is critical, the more expensive series 525 possesses the same physical properties as the standard series.

The most immediately obvious advantage of structural fiberglass is its relative weight: it is almost 80% lighter than steel and over 30% lighter than aluminum. As one of our major concerns often centers on making scenery as lightweight as possible, this material offers a most favorable alternative.

Structural fiberglass was developed originally for use in corrosive environments; therefore, it will not rot or absorb moisture. It cannot rust, and it provides excellent electrical insulation, due to its non-conductive properties. These factors combine to make it an ideal choice for outdoor structures, as well as more traditional scenic uses.

The cost of structural fiberglass falls somewhere between steel and aluminum for most shapes and sizes. In several instances, such as small angles and wide-flange I-beams, the fiberglass alternatives are slightly more expensive than aluminum. Round tube, however, is consistently less expensive than either steel or aluminum.

The single most important advantage in working with structural fiberglass is that it requires no special tools. Construction can be accomplished with a hacksaw, drill, and all other standard woodshop tools. The only necessary precaution regarding power tools is to be certain that the fiberglass dust is cleaned out after use, since the abrasive glass particles could damage moving parts.

Fiberglass shapes can be joined to themselves or other materials with mechanical or adhesive connections. For permanent joints, a combination of methods will provide the greatest strength. Depending on the intended use, reusability concerns, and the sections to be joined, joints can be fastened by bolts, screws, rivets, nails, epoxies, or panel adhesives. Obviously, as with any connection, the amount and type of stress to be placed on the joint must be considered.

Latex, vinyl, and casein paints are all compatible with structural fiberglass. Use of a universal colorant in polyester resin will produce a very hard and shiny finish. Additionally, most common texturing materials adhere securely to the surface of the fiberglass.

The only significant disadvantage of fiberglass is that it cannot be bent and is not, therefore, appropriate for curved surfaces. Although this limits its usefulness, structural fiberglass is often ideal for straight line work.

As with any material, certain safety precautions must be observed. Safety goggles and dust masks should be worn when cutting, drilling, sanding, or machining structural fiberglass because the dust produced can easily damage the eyes, nose, and lungs. To protect the skin, latex or vinyl gloves, long-sleeved shirts, and jeans are recommended. The gloves will also protect your hands from most of the chemicals used in connection with fiberglass resins. Whenever your work involves the use of polyester resins, which contain styrene and methyl ethyl ketone peroxide, a NIOSH-approved organic vapor respirator should be worn with a dust/mist prefilter attached. Respirators are absolutely vital where there is inadequate ventilation — a condition existing in many scene shops. A few relatively simple safety precautions will insure the health of all persons working in the shop while the structural fiberglass is in use.

Structural Fiberglass and Its Applications

Andi Lyons

I have found structural fiberglass extremely useful in a variety of applications. The following describes two of the many uses to which we have put fiberglass shapes in the past several years.

Our mainstage floor has several significant warps that lift parts of wagons off the floor at various points in their travel. We have been faced with the problem of tracking wagons directly on our stage floor, so that they will run as precisely and quietly as possible. For one production, fiberglass angle fins were substituted for the more common steel fins on three of six wagons, which varied in height from 8" to 3'-3", and in footprint from 8' x 8' to 12' x 24'. The fiberglass fins were set at the appropriate angle for each wagon and bolted into place, as were their steel counterparts. Each fin was as long as possible to provide the maximum amount of contact with the plywood tracks. The fiberglass fins were noticeably quieter along the tracks. The use of wax in the tracks further improved the ride, as did treating the tracks with a silicone spray. The fiberglass fins remained significantly less noisy than the steel fins of the control group throughout the run of the production. As a result, since the test, we have used fiberglass angles for all our wagons.

The director of another production requested two firepoles for actor entrances. In searching for the most cost-effective response to that request, we found an 18'-long 3"-diameter aluminum tube priced at $115.38 — and same-size tube in structural fiberglass at $86.58. Considering the savings, we opted for the fiberglass. A black universal colorant added to polyester resin and painted onto each pole provided the shiny black finish the designer required. The technique offered the bonus of being especially resistant to chipping and abrasion, and actually made the poles easier to slide down while providing the desired high-gloss look.

Though structural fiberglass may not be the panacea for all theatrical construction problems, it does merit serious consideration. Although it cannot be bent or reshaped, the simplicity of construction and its excellent strength-to-weight ratio give it the potential for becoming as popular as aluminum and steel for many applications. To date, I have used fiberglass I-beams as carrying members for spans, square tube for platforms and flats, round tube for columns and beams, and angle for flats and wagon fins. It appears that for times when weight is a critical factor, for outdoor events, for wet situations such as ice shows, and for a variety of other usual and unusual problems, structural fiberglass shapes may well offer workable solutions in terms of time and money.

MAJOR DISTRIBUTORS OF STRUCTURAL FIBERGLASS

Ain Plastics, Inc.
249 East Sandford Blvd.
Mount Vernon, NY 10550
(914) 668-6800

Joseph T. Ryerson & Son, Inc.
Plastics Division
(Check Yellow Pages for nearest service center.)

TECHNICAL BRIEF

Scenery Mechanics

Since hydraulic systems excel in moving heavy objects with virtually no noise, they can be found driving lifts, wagons, and turntables in many theatres. A small hydraulic system is not too difficult to assemble, and will find repeated use. While a full description of every part in a hydraulic system would be very long, a brief overview can provide a useful amount of information about these systems. Every hydraulic system will contain components from each of the six categories described below: Pumps, Actuators, Valves, Fluid Conductors, Fluid, Storage.

PUMPS

Pumps convert the mechanical power provided by an electric motor into hydraulic power. They are rated in terms of nominal flow rate, maximum operating pressure, and input shaft speed. Manufacturers supply pressure/volume curves that express the characteristics of a given pump.

Many different types of pumps are available, but for theatre — where price and simplicity are more important than long-term efficiency — gear or vane type pumps are most common. The hydraulic power provided by a pump can be calculated from:

$$W = \frac{QP}{1714}$$

where:
W = power (hp)
Q = flow rate (gpm)
P = pressure (psi)

As a guideline, the smallest capacity pump that should be considered for general purpose use in theatre would be one that puts out about 7gpm at 800psi, or about 3hp. Pumps create high levels of noise and must be isolated from the stage in a distant room or surrounded with a soundproof box.

ACTUATORS

Actuators convert fluid power to mechanical power. For linear motion, the hydraulic cylinder is elegantly simple. A long rod attaches to a piston that moves inside a tube. Ports at either end of the tube allow oil to flow in or out. Cylinders are rated in terms of bore (the diameter of the piston), stroke (how far the piston moves), rod size, and maximum operating pressure. The pressure supplied by the pump acts on the piston area to create a force. The pulling and pushing forces differ because the input pressure acts on the full area of the piston when it is extending the rod, but while retracting, the pressure acts on only that part of the piston not covered by the rod.

$$F_{push} = PA_{piston}$$
$$F_{pull} = P(A_{piston} - A_{rod})$$

where:
F_{push} = pushing force (lbs.)
F_{pull} = pulling force (lbs.)
P = pressure (psi)
A_{piston} = area of the piston (sq. in.)
A_{rod} = area of the rod (sq. in.)

Cylinder speed is a function of flow rate and cylinder dimensions:

$$V = \frac{.3208Q}{A}$$

where:
V = speed of the rod end (ft./sec.)
Q = flow rate (gpm)
A = area of piston during extension or area of piston minus area of rod during retraction (sq. in.)

For example, assume a pump provides 10gpm at 800psi at a 2.5"-bore, 1"-rod, 2'-stroke cylinder. On extension, the cylinder could provide:

$F_{push} = PA_{piston}$

$F_{push} = 800psi \times (\pi(1.25\ in.)^2) = 3927\ lbs.$

$$V = \frac{.3208Q}{A}$$

$$V = \frac{.3208\ (10gpm)}{\pi(1.25\ in.)^2} = 0.654\ ft./sec.$$

Hydraulic motors are also available to provide rotary output, but their complexity precludes further discussion here.

VALVES

Valves control the direction, volume, pressure, and distribution of flow. Only the relief valve and 4-way valve are worth mentioning in this overview. Many types of pumps have no inherent limit to their output pressure, though manufacturers' specifications require that a rated maximum pressure never be exceeded. A relief valve will keep the output pressure of a pump from exceeding some set value by venting oil back to the reservoir when the pump pressure begins to exceed the relief pressure. This valve is built into some pumps.

The 4-way valve is the most common device used to control the movement of actuators. The "way" refers to the number of ports on the valve: one high-pressure inlet, one return to the reservoir, and two ports that go to the "in" and "out" of a motor or one to each end of a cylinder. These ports usually come with female threads in standard nominal pipe sizes.

Valves are rated by the maximum pressure they can withstand and the pressure drop across them as a given amount of oil flows through.

A 2-position valve will not allow a cylinder to be stopped mid-stroke or a motor to be stopped at all, so 3-position valves are most common in theatre. In their center position, these "closed-center" valves prevent the movement of fluid and, hence, any movement of the actuator. Some 3-position valves have detents that hold them open in forward, reverse, or neutral, while others are spring-centered and act as dead-man controls.

Four-way valves can be controlled manually (with hand levers, pushbuttons, or footpedals), electrically (with solenoids), or by hydraulic or pneumatic pressures fed to them from other valves.

A representative full specification for a typical valve would therefore be a "4-way, 3-position, closed center, $\frac{1}{2}$"-port, hand-lever-actuated, spring-centered valve."

FLUID CONDUCTORS

Fluid conductors convey fluid from one part of a system to another. This plumbing is usually done with standard steel pipe, rubber hydraulic hose, or steel tubing. Pipe is best for long, straight, permanent runs; hose, for hookup to valves and actuators in temporary setups; and tubing, for intricate bends and connections in locations that offer protection from physical abuse. It is very important to size conductors so that a minimal amount of system pressure is lost pushing oil through the plumbing. For typical theatre systems, it is best to limit the speed of oil through the plumbing to 15 feet per second (fps), thereby limiting pressure loss.

The formula below is used to determine plumbing diameter given this 15fps limit.

$$\mathbf{d} = \sqrt{.0272\mathbf{Q}}$$ where: **d** = the fluid conductor's inside diameter (in.)
                                                          Q = flow rate (gpm)

What size conductor would be needed to adequately deal with a flow rate of 15gpm?

$$\mathbf{d} = \sqrt{.0272(15\text{gpm})} = 0.638 \text{ in.}$$

Pipe comes in nominal sizes with varying inner diameters depending on the schedule. Hose is sized by inner diameter; and tubing, by outer diameter and wall thickness. Therefore, a 15gpm flow would efficiently pass through a $\frac{1}{2}$" schedule 40 black pipe, a $\frac{3}{4}$" hose, or $\frac{3}{4}$" x 0.058" wall tubing. This simplified approach to conductor sizing works well if plumbing runs do not exceed about 50 feet.

FLUID

The fluid in a hydraulic system is usually oil. Oils can be refined to different thicknesses, or viscosities. Pump specs determine what viscosity oil should be used in a system, but a commonly used oil viscosity is 200 SSU (Saybolt Seconds Universal), which is similar in thickness to 10-weight motor oil. Motor oil should not be used in hydraulic systems because it contains additives incompatible with the seals used in most hydraulic devices.

The viscosity and chemical makeup of an oil can change significantly when heated excessively, so the oil must be kept cool. Above 140° F, hydraulic oil quickly undergoes permanent damage.

STORAGE

Storage of oil in reservoirs is essential for cooling, and for supplying oil to cylinders. (The volume of oil in a fully extended cylinder is greater than in a fully retracted one.) A reservoir should have a capacity at least 3 times the flow rate of the pump, *e.g.*, a 15gpm pump should sit over a 45-gallon reservoir. Conditioning equipment includes filters that clean the oil and oil coolers.

Assembling a hydraulic system to meet exact performance specifications is extremely difficult. Hydraulic systems are relatively inefficient, and losses can crop up in many places not easily predicted without a thorough engineering analysis. It is always best to design systems to perform 30% to 50% above known needs.

REFERENCE

Industrial Fluid Power, Volumes 1, 2, & 3
Womack Educational Publications
P.O. Box 35027
Dallas, TX 75235

Air casters powered by a large volume of low-pressure air from a vacuum cleaner can be constructed in any woodworking shop. Most vacuum cleaners will supply sufficient air for two 12"-radius casters, and a typical system using two vacuum cleaners and four 12"-radius casters can move over one ton. The following design and construction procedures work for casters with radii between 9" and 12". Sealing the gaskets of larger casters becomes difficult.

DESIGN PROCEDURE

1. Determine the pressure generated by a given vacuum cleaner with the following formula:

    $P = W/\pi r^2$ where: P = pressure (psi)
                                              W = maximum weight that can be lifted (lbs.)
                                              r = radius of the vacuum cleaner nozzle (in.)

    Find W by holding the nozzle vertically, covering it with a piece of $\frac{1}{4}$" plywood, and adding weight until air can no longer escape.

2. Calculate the radius of the air casters needed by applying the following formula:

    $R = W\sqrt{L/.9P\pi}$ where: R = required radius of the air casters (in.)
                                              L = load to be supported by one caster (lbs.)
                                              P = pressure supplied by the vacuum cleaner (psi)

CONSTRUCTION PROCEDURE

1. The most sophisticated part of the air caster is the gasket, which is cut from an inner tube. The outside radius of the inner tube should be equal to R when inflated but not stretched. The inside radius should be 4" less than R. One inner tube is required for each air caster. See Figure 1 for instructions on cutting the gasket from the tube.

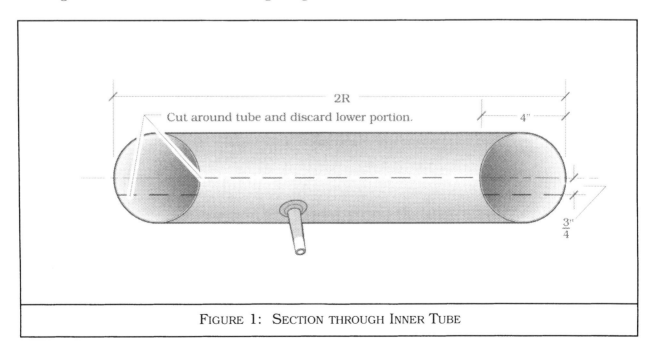

FIGURE 1: SECTION THROUGH INNER TUBE

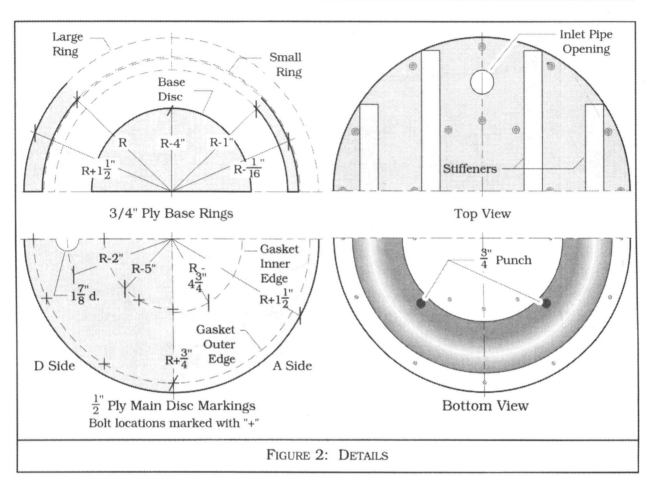

3/4" Ply Base Rings

Large Ring

Small Ring

Base Disc

R

R-4"

R-1"

R+1½"

R-1/16"

Inlet Pipe Opening

Stiffeners

Top View

Gasket Inner Edge

R-2"

R-5"

R-3¾"

R+1½"

1⅞" d.

Gasket Outer Edge

R+¾"

D Side

A Side

½" Ply Main Disc Markings
Bolt locations marked with "+"

¾ Punch

Bottom View

FIGURE 2: DETAILS

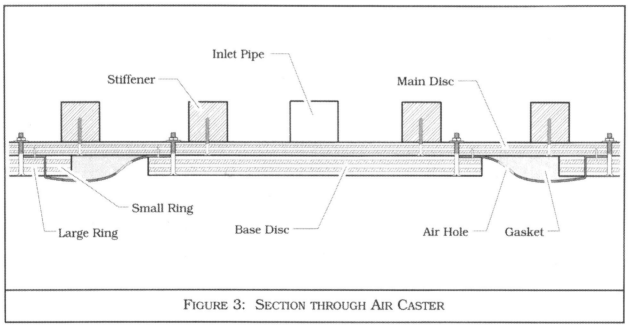

Inlet Pipe

Stiffener

Main Disc

Small Ring

Large Ring

Base Disc

Air Hole

Gasket

FIGURE 3: SECTION THROUGH AIR CASTER

2. Cut the main disc and mark the location of all components on it. The 2x2 stiffeners and the bolt locations are marked on the "D" side of the plywood. The stiffeners should be perpendicular to the plywood grain. Draw the gasket lines, on the "A" side. See Figure 2.

3. Cut the hole for the inlet pipe, which will be press-fitted after the caster is complete.

4. Glue and screw the stiffeners to the main disc.

5. Staple the outer circumference of the gasket to the bottom of the main disc on 1" centers.

6. Place the large ring over the gasket and bolt the ring to the main disc. Use twelve 2" x $\frac{3}{16}$" flat head stove bolts. Countersink the bolts. See Figure 3.

7. Install the small ring inside the gasket, pressed against the outer ring. To install the ring, cut it in half, then cut an additional $\frac{3}{8}$" from one section. Attach the halves one at a time using glue and nails.

8. Staple the inner circumference of the gasket on 1" centers.

9. Attach the base disc as in step 6 above, using eight 2" x $\frac{3}{16}$" bolts.

10. Punch four $\frac{3}{4}$" air holes in the gasket. The holes should be one inch from the base disc, evenly spaced around the gasket, and offset from the inlet pipe.

11. Press-fit the inlet pipe into the caster until it is flush with the bottom of the main disc, then seal with a bead of glue. The vacuum cleaner can be connected to the caster with 2" flexible hosing and PVC connectors.

With all air casters, a smooth, non-porous floor is required for proper operation. Also, the noise of the vacuum cleaners in this system may make it impractical for some uses.

❧❧❧❧

Shop-Built Pneumatic Cylinders

Light-duty pneumatic cylinders can be constructed in almost any shop. They are economical and relatively maintenance free, and they provide a push or pull force that can be remotely controlled. All materials can be purchased from plumbing and steel supply companies. The 6'-0"-long 2"-diameter cylinder described here costs about $100.00. Smaller or larger cylinders are possible.

PRINCIPLE OF PNEUMATIC CYLINDERS

See Figure 1. A rod connects the piston and the work to be moved. A pressurized gas (air, CO_2, or nitrogen) forces a piston through a cylinder. The greater the pressure, the greater the force that moves the piston. The force (lbs.) can be calculated by multiplying the surface area (sq. in.) being acted upon by the air pressure (psi).

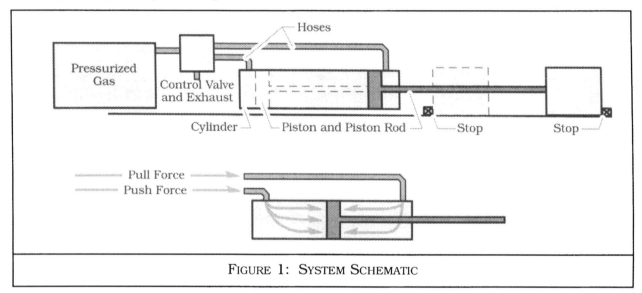

FIGURE 1: SYSTEM SCHEMATIC

THE SYSTEM

The system is controlled by valves that connect the cylinder to the pressurized air supply and the exhaust. Air forced in one side moves the piston and forces air out the other side to the exhaust. When the pressures are reversed, the piston moves in the opposite direction. Given constant air pressure, a pull force is smaller than the push force because the rod takes up a certain amount of surface area on the pull side. See Figure 1. The cylinder should be securely mounted, and the work should have built-in stops to limit its movements. Lubricate the system periodically.

CAUTION

Much of the safety and dependability of this device depends on accurate shop procedures. Be sure to observe the following guidelines for safe work:

1. Test the cylinder before installation: start with low pressures and work up.

2. The cylinder can explode if too great a pressure is used, or if the cement doesn't bond properly. Do not exceed 100psi of pressure.

3. Never stand in front of either end of the cylinder.

4. Never use compressed oxygen.

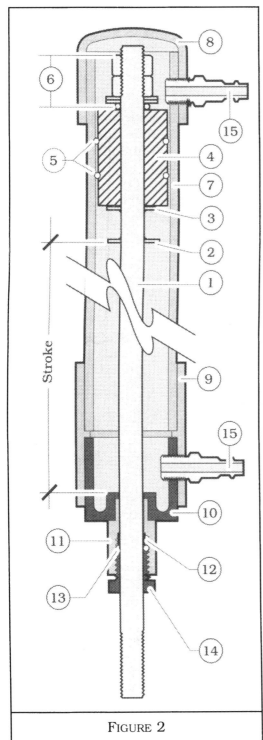

FIGURE 2

2"-DIAMETER PNEUMATIC CYLINDER PARTS LIST AND
CONSTRUCTION NOTES — SEE FIGURE 2

1. Rod: $\frac{1}{2}$" cold-rolled steel, threaded at both ends —
    (Length = stroke + approximately 8")

2. Stop Washer: $\frac{3}{8}$" washer drilled to $\frac{1}{2}$" I.D. — brazed
    to rod, 1" from part 3

3. Retaining Washer: $\frac{3}{8}$" washer drilled to $\frac{1}{2}$" I.D. —
    brazed to rod; seal between parts 3 and 4 with sili-
    cone caulking

4. Piston: birch or oak, 2" long — lathe-turned for
    loose fit in cylinder, with grooves for O-rings and
    sealed with Varathane

5. Two O-rings: $\frac{3}{16}$" cross section, $1\frac{5}{8}$" I.D.

6. Forward Retainer Assembly:

    O-ring: $\frac{1}{8}$" X-section, $\frac{1}{2}$" I.D.
    $\frac{3}{8}$" washer drilled to $\frac{1}{2}$" I.D.
    $\frac{1}{2}$" lock washer
    Two $\frac{1}{2}$" nuts

7. Cylinder: 2" schedule 40 PVC (280 psi)

8. 2" schedule 40 PVC end cap

9. 2" schedule 40 PVC coupler

10. 2" to $\frac{1}{2}$" SS PVC bushing

11. $\frac{1}{2}$" ST PVC adapter

12. $\frac{1}{2}$" EMT (conduit): $\frac{1}{4}$" long — cut with pipe cutter to
    flare ends in

13. O-ring: $\frac{1}{8}$" cross section, $\frac{1}{2}$" I.D.

14. $\frac{1}{2}$" to $\frac{3}{8}$" PVC bushing

15. Two $\frac{1}{4}$" male air quick connects (NPTM).

ASSEMBLY PLAN

First, assemble parts 1 through 6. Lubricate the cylinder
(with silicone-based O-ring lube only). Test that the O-
rings maintain an airtight fit in the cylinder.

Second, assemble parts 7 through 11 using Weld-On P-70®
Primer and Weld-On Cement #705®.

Pneumatic Door Stabilizer

Jeff Dennstaedt

THE PROBLEM

Doors (particularly lightweight doors) that bounce open when they are slammed.

THE DESIGN GOALS

Develop a door-stabilizing mechanism that will allow a door to be closed gently or slammed shut and remain closed; work on all types of doors, including lightweight doors; automatically reset itself; and be visually acceptable, quiet, and reliable.

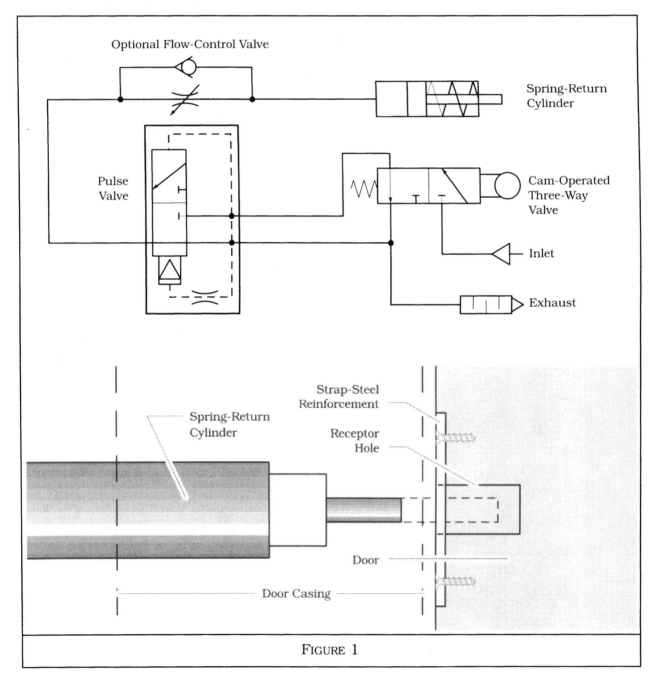

FIGURE 1

THE SOLUTION PRINCIPLE

The impact of slamming a door is often too great to be absorbed by a normal door latch or magnetic catch. The result is that the door bounces back open. Our solution to this problem includes a magnetic door catch to hold the door in place — not to stop it. To stop the door, we are using a miniature pneumatic cylinder (plumbed as shown in Figure 1) whose piston extends into the door when the door is slammed. Once the door is stabilized by the cylinder's piston rod, the magnetic catch holds the door. The piston rod automatically retracts half a second later.

SYSTEM OPERATION

The key to the system's operation is a special one-shot pulse valve manufactured by Fabco-Air. When pressure is applied to the valve's inlet, a short pulse of air (one-half to three-quarters of a second) is released. The valve then shuts off and will not fire again until the system pressure has been removed and reapplied. In this way the door can begin closed, with the single-acting, spring-return cylinder's piston retracted, at which time the 3-way normally closed valve is open. As the door is opened, the 3-way valve returns to its closed position, stopping pressure to the pulse valve and allowing the pressure in it to be exhausted, thus resetting the valve. As the door is closed, the 3-way valve is opened and sends a short pulse of air to the cylinder, causing the piston rod to extend into the steel-reinforced edge of the door, stabilizing the door. At the end of the pulse, pressure stops flowing to the cylinder, the pressure in the cylinder is exhausted, the internal piston spring retracts the rod into the cylinder, and the entire system is restored to its initial state.

COMMENTS ON SYSTEM OPERATION

In trying to make the mechanism suitable for the widest range of closing forces, we discovered the extension speed of the piston to be a limiting factor. Regardless of the pressure, a very slight but potentially significant time delay occurs between the 3-way valve's activation and the extension of the piston rod. A delay of only a few milliseconds is enough to allow the door to travel some distance. When the door is closed gently it travels only a slight distance. However, when the door is slammed, it can travel a considerable distance before the piston can fully extend. In fact, in some tests it was possible to slam the door hard enough that the door activated the valve, hit the door stop, and bounced back beyond the piston before the piston could extend into the door.

In order to remedy this situation, we increased the extension speed by switching to a smaller diameter cylinder, thus reducing the chamber volume to be filled. This helped significantly, but some experimentation with the placement of the piston relative to the activation point of the 3-way valve was necessary to accommodate a wide range of closing forces. To accommodate a greater range would require larger valving and tubing to increase the flow rate to the cylinder and thus increase the piston speed. Clippard Minimatic manufactures smaller diameter cylinders, but we would caution choosing any smaller than ours because the piston rod is subjected to side loading, and rods of less than $\frac{1}{4}$" diameter may become bent and unusable.

PARTS AND FITTINGS

Except as noted, all of the parts and fittings below are manufactured by Clippard Minimatic.

Part Description	Part Number	Quan	Unit Price	Extended Price
3-Way Valve	FV-3	1	8.80	8.80
Flow-Control Valve	MVC-2	1	5.45	5.45
Pulse Valve[1]	OS-1	1	17.90	17.90
Muffler	15070	1	1.55	1.55
1" Spring-Return Cylinder	H9S-1S	1	11.65	11.65
Mini-Cam Follower	11925	1	1.70	1.70
Cylinder Mounting Bracket	15018-2	1	1.05	1.05
$\frac{1}{4}$" FPT Female Plug[2]	2X170	1	.92	.92
$\frac{1}{4}$" NPT to 10/32	15006-3	1	3.80	3.80
$\frac{1}{8}$" NPT to $\frac{1}{8}$" Hose Barb	11924-1	3	1.06	3.20
10/32 to $\frac{1}{8}$" Hose Barb	11752-3	9	.31	2.80
X-Fitting	15002-4	1	4.90	4.90
Female Hex Connector	15004	1	2.05	2.05
Screw Plug	11755	1	1.20	1.20
$\frac{1}{8}$" NPT to 10/32	15006-1	1	2.90	2.90
Gasket	11761-2	1	2.40	2.40
$\frac{1}{8}$" ID hose (50')	3814-1	1	7.00	7.00
				77.27

[1]Fabco-Air, Inc.
[2]W. W. Grainger, Inc.

SUPPLIERS

Clippard Inst. Lab., Inc.
7390 Colerain Rd.
Cincinnati, OH 45239
(513) 521-4261

Fabco-Air, Inc.
3718 N.E. 49th Rd.
Gainesville, FL 32601
(904) 373-3578

W. W. Grainger, Inc.
Distribution Group General Office
5959 West Howard St.
Chicago, IL 60648
(312) 647-8900

A problem often faced in scene shifts is how to lock rolling scenery in its playing position quickly. Rather than using drop pins for a Pennsylvania Center Stage production of *Peter Pan*, we devised the lift jack described below, which raised the back casters off the deck, thereby preventing the wagons from moving and eliminating the problem of finding holes in the dark.

This particular design proved itself superior to more familiar lift jacks in a number of ways. The simplicity of the design makes it exceptionally easy and inexpensive to construct. The ease of operation makes it quick and very efficient. And most importantly, its reusability makes it a convenient device to have on hand as stock hardware.

MATERIALS

> 6" piece of $1\frac{1}{4}$" schedule 40 black pipe
> 4" piece of 2" schedule 40 black pipe
> 5" piece of $\frac{1}{4}$" x $1\frac{1}{2}$" piece of flat stock
> $\frac{3}{8}$" x 6" hex head bolt

DESCRIPTION

The lift jack, shown in Figure 1, is a 6"-long piece of $1\frac{1}{4}$" schedule 40 black pipe sleeved into a 4"-long piece of 2" schedule 40 black pipe. Using a metal chop saw, we cut a 30° slot at the midway point in the 2" pipe, wide enough to accommodate the shaft of a $\frac{3}{8}$" x 6" hex head bolt. At each end of the slot we filed a small detent so that the bolt would lock the jack in either the playing or traveling position. We next drilled and tapped a series of $\frac{3}{8}$" holes along the side of the $1\frac{1}{4}$" pipe spaced at $\frac{1}{2}$" o.c. to give the jack $\frac{1}{2}$"- increment height adjustment. At the top of the 2" pipe, we welded a $\frac{1}{4}$" x $1\frac{1}{2}$" x 5" plate pre-drilled for four $\frac{3}{8}$" bolts for mounting to the platform. The $\frac{3}{8}$" x 6" hex head bolt which we later inserted into one of the tapped holes served as a handle and held the jack assembly together. If there is difficulty in positioning the lever arm a piece of 1" pipe can be fitted over the end of the hex head bolt to provide any necessary mechanical advantage.

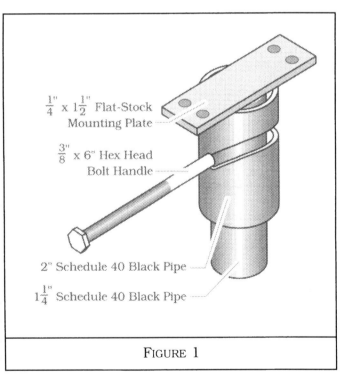

$\frac{1}{4}$" x $1\frac{1}{2}$" Flat-Stock Mounting Plate

$\frac{3}{8}$" x 6" Hex Head Bolt Handle

2" Schedule 40 Black Pipe

$1\frac{1}{4}$" Schedule 40 Black Pipe

FIGURE 1

SOURCE

Tim Irish.

TECHNICAL BRIEF

Scenery Tools

Styrofoam® Moulding Cutter

Rob Chase

Pictured below is a wood lathe attachment for producing Styrofoam® moulding. This low-cost attachment is easy to construct and can be adapted for a variety of moulding shapes and sizes. It consists of three basic components: cutter, cutter housing, and clamping board. See Figure 1. Like any router blade, the cutter is a double-faced "negative" of the moulding it will produce. The cutter housing both guides the Styrofoam® past the cutter and collects the chips produced by its operation. The clamping board holds the attachment securely in place on the lathe bed.

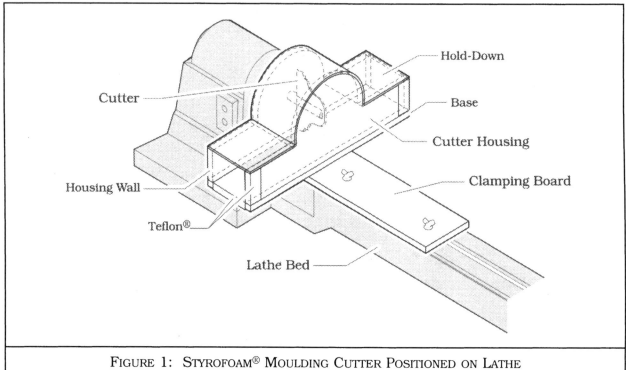

FIGURE 1: STYROFOAM® MOULDING CUTTER POSITIONED ON LATHE

CONSTRUCTING THE CUTTER

1. To make a pattern for the cutter, fold a piece of paper in half. Draw a line parallel to the fold and $\frac{1}{2}$" away from it. See Figure 2. Working between that line and the edge of the paper, lay out the moulding profile you wish to produce. At the ends of the moulding profile line, draw lines perpendicular to the fold and to the edge of the paper. The length of the line perpendicular to the fold equals moulding depth (**md**) plus $\frac{1}{2}$". The length of the line perpendicular to the edge of the paper equals moulding height (**mh**).

2. Trace the pattern's outline on a piece of $\frac{1}{4}$" plywood. Draw a centerline along the resulting shape's axis of symmetry and then draw a pair of lines parallel to the centerline and $\frac{1}{2}$" away from it. Cut along the outline of the cutter. Bevel the leading edge 30°. See Figure 3.

3. Notch a 1" hardwood dowel (the cutter shaft) as deep as the cutter's shape will allow without letting the dowel interfere with the cutting action.

4. Center the dowel on the centerline (see Step 2) and drill two $\frac{3}{16}$" holes through the shaft and cutter. Bolt the shaft and cutter together with $\frac{3}{16}$" x $1\frac{1}{4}$" round head stove bolts.

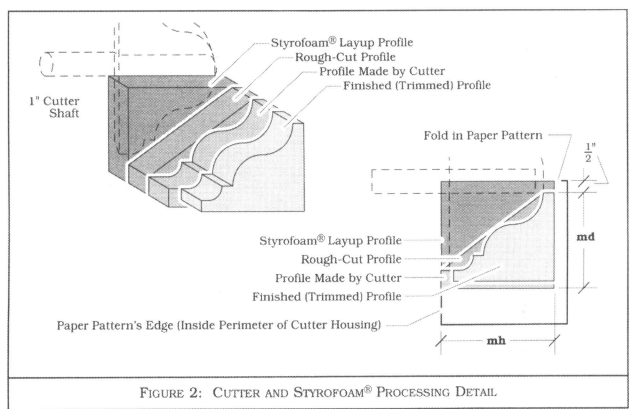

FIGURE 2: CUTTER AND STYROFOAM® PROCESSING DETAIL

CONSTRUCTING THE HOUSING

1. To make the housing base, rip a 2'-0" long piece of $\frac{3}{4}$" plywood to $1\frac{9}{16}$" wider than **mh**. See Figure 2. Glue and staple a clamping board of $\frac{3}{4}$" plywood 18" long to the underside and at right angles to the base, and centered on it. On the clamping board centerline, drill two $\frac{5}{16}$" holes about 12" apart. The clamping board will later be bolted to the lathe bed with two $\frac{5}{16}$" hex head bolts, each of which passes through a piece of $\frac{1}{4}$" bar stock long enough to span the bottom of the lathe bed.

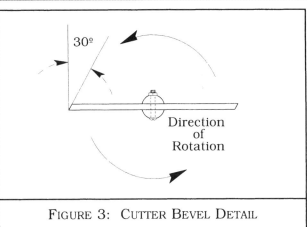

FIGURE 3: CUTTER BEVEL DETAIL

2. Cut two pieces of $\frac{3}{4}$" plywood to 2'-0" for the housing walls. Draw a line on the face of each piece at a distance of moulding depth **md** from a 2'-0" side. At the center of the moulding depth line on each piece, draw an arch with a radius of **md** + $1\frac{1}{2}$" at the center of the wall height line. See Figure 2. Cut out the walls.

3. In one of the walls cut a $1\frac{1}{8}$"-diameter hole at the center of the wall height line to allow for insertion of the cutter shaft through the wall and into the lathe chuck.

4. Glue and staple the housing walls to the pre-assembled base and clamping board, being careful to keep the distance between the walls **mh** + $\frac{1}{16}$". See Figure 2.

5. Cut two strips of $\frac{1}{16}$"-thick Teflon® 2'-3" long. Make one of them wide enough to cover the inside of the housing base; the other, wide enough to cover the inside of the housing wall. Place these pieces in position and staple the ends to the outside of the housing. See Figure 1.

6. Clamp the clamping board and housing in position on the lathe bed and secure the cutter in the chuck.

7. Cut two pieces of Masonite® at least 6" long and wide enough to span the housing walls. Bevel one end of each piece and then attach these hold-downs on top of the housing, butted to the arch with the beveled edges toward the in-feed end.

8. Cut a housing cover out of a piece of cardboard wide enough to span the housing walls and long enough to cover the arch. If you plan to attach the cutter to a shop-vac, cut a hole to receive the shop-vac hose in the in-feed side of the housing cover.

FINE-TUNING THE CUTTER

1 Rough-cut the Styrofoam® stock on the bandsaw to approximate the finished moulding shape.

2. Make sure the housing is tightly clamped to the lathe bed. Hand turn the spindle to check for clearance. Turn on the lathe at a low RPM and feed the Styrofoam® into the housing, cutting several inches to shape. Turn off the lathe.

3. Without removing the cutter from the chuck, compare each cutter edge to the finished cut and sand the cutter edges as needed.

4. When the cutter has been fine-tuned, place the housing cover in place over the arch, and tape the optional shop-vac hose in place.

5. Turn the spindle by hand to check clearance. Turn on the lathe at a low RPM. Experiment with RPM and feed speed to achieve the best cut.

6. After you have shaped the moulding face, cut the moulding to the desired depth and height.

ɜ▲ɜ▲ɜ▲

Roto Locks® or "coffin locks" are common stage hardware used for locking platform decks together. Presented here is a shop-built jig used to install Roto Locks® quickly in either of two differently constructed decks. In both cases the locking hardware is installed after the deck has been built.

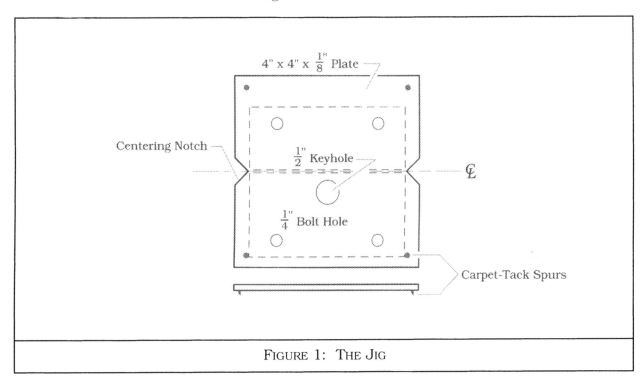

FIGURE 1: THE JIG

THE JIG

The jig is constructed of a piece of 4" x 4" x $\frac{1}{8}$" steel with four $\frac{1}{4}$" and one $\frac{1}{2}$" hole drilled into it. These holes correspond to the Roto Locks'® four bolt holes and keyhole. In addition, a small notch is cut in each side of the jig at the center. These notches will serve to locate the jig accurately when drilling into the platform lid. It is important to locate each half of a Roto Lock® $\frac{1}{16}$" from the edge of the platform deck so that the Roto Lock® can pull the platforms together tightly. To build this $\frac{1}{8}$" gap into the jig, mark the centers of the keyhole and the bolt holes for each half of the Roto Lock® $\frac{1}{16}$" farther from the centerline of the 4" plate before drilling the holes in the jig.

To insure that the jig can be held firmly in place while drilling into the deck, drill four additional holes, one at each corner of the jig, just large enough to accept a $\frac{1}{4}$" carpet tack. Spot-weld these tacks into the jig so they will act as spurs that hold the jig in place.

INSTALLATION IN STANDARD CONSTRUCTION PLATFORMS

For platforms constructed with 1x6 stringers on the outside edge, use the jig to install the Roto Locks® in the following manner. See Figure 2.

1.   With the pre-built deck assembled on the shop floor, place the jig on the deck at the location of each Roto Lock®. Drill through all 5 holes with a $\frac{1}{4}$" drill bit. Repeat this process for all the Roto Locks® being installed, taking care not to place a Roto Lock® where there is either an interior stringer or a nail in the lid.

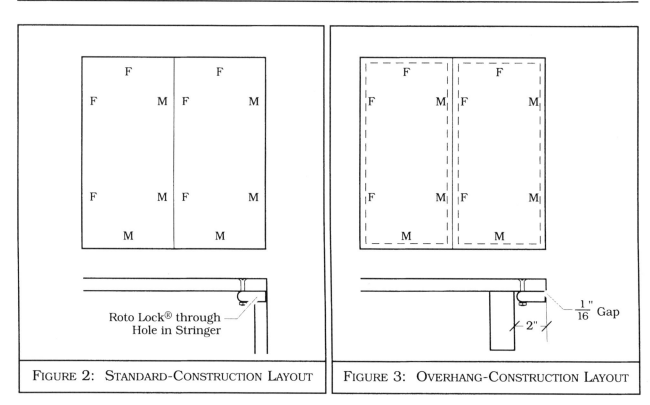

FIGURE 2: STANDARD-CONSTRUCTION LAYOUT   FIGURE 3: OVERHANG-CONSTRUCTION LAYOUT

2. Lift each platform onto sawhorses and enlarge the keyhole with a $\frac{1}{2}$" twist drill bit.

3. Use the $\frac{1}{2}$" twist drill bit to drill a hole in the side stringer at each Roto Lock's® location. Enlarge the hole in the stringer with a saber saw to accept a Roto Lock®.

4. Install the Roto Locks® underneath the lid using $\frac{1}{4}$" flat head stove bolts.

5. Do a trial assembly after all the platforms have been equipped with Roto Locks®.

INSTALLATION REQUIRING SPECIAL PLATFORM CONSTRUCTION

Build each platform with a 2" overhang on all four sides. See Figure 3. Locate the Roto Locks® along the platforms' edges in a uniform fashion so that the platforms remain interchangeable. This approach works well on stressed-skin panels and is ideal for making stock platforms.

1. Laying each platform in turn on a work table, mark each platform where the Roto Locks® are to be placed. Note on the platform the location of male and female Roto Lock® halves.

2. Place the appropriate half of the jig on the platform and drill the holes.

3. Install the Roto Locks® using $\frac{1}{4}$" flat head stove bolts.

4. Do a trial assembly.

❧❧❧❧

$\frac{1}{2}$" Homasote® which has been sheared in two along its edge is an attractive alternative to $\frac{1}{4}$" plywood as a material for use in making brick facings for the stage. For one thing, each 4x8 sheet yields up to 64 square feet of bricks, making the cost quite attractive. For another, the sheared surface requires little further treatment to achieve a rough, brick-like quality. Finally, installation can be quickly and easily accomplished by use of a matte knife, glue, and a stapler.

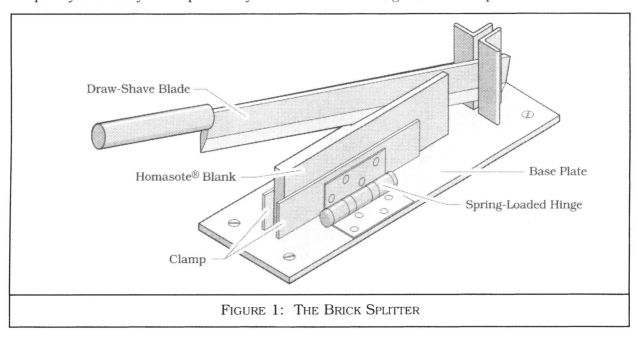

FIGURE 1: THE BRICK SPLITTER

Figure 1 illustrates a shop-built tool for splitting brick-size Homasote® blanks into finished brick facings. Essentially, the splitter consists of a clamp, a blade, and a base plate. During operation, the user inserts each blank in turn into the clamp, forces the blade through the blank, and then removes the finished pair of bricks. Few if any of the dimensions or materials to be used in making the brick tool are truly essential, but a few suggestions are in order.

1.   Providing mounting holes in the corners of the steel base plate insures stability during its use.

2.   The use of a spring-loaded hinge on one side of the clamp allows for ease of Homasote® insertion and removal and also eases the passage of the blade through the Homasote®.

3.   The distance between the sides of the clamp should be a little greater than $\frac{1}{2}$" at the base plate, though the hinged side of the clamp should be designed to lean in toward the blade to insure a snug fit around the uncut blank.

4.   A draw-shave makes a fairly good blade, though a blade made of sharpened $\frac{3}{16}$" flat steel and fitted with a handle would probably also work.

5.   The blade's pivot should be located in such a manner above the base plate as to allow the entire edge of the blade to come into contact with the base plate at the completion of each pass.

Using this splitter, a carpenter can quickly turn out inexpensive brick facings that are very easy to attach to hard-covered flats and platforms and that have the advantage of requiring much less finishing time in the hands of the paint shop.

A Pantograph Moulding Jig

Jeff Dennstaedt

Usually, the process of mitering mouldings requires measuring an angle, dividing it by two, and setting a saw to the desired cutting angle. This time-consuming task can be accomplished more quickly and accurately with the shop-built jig shown in Figure 1. This pantograph jig transfers angles by means of a set of interconnected, pivoting arms.

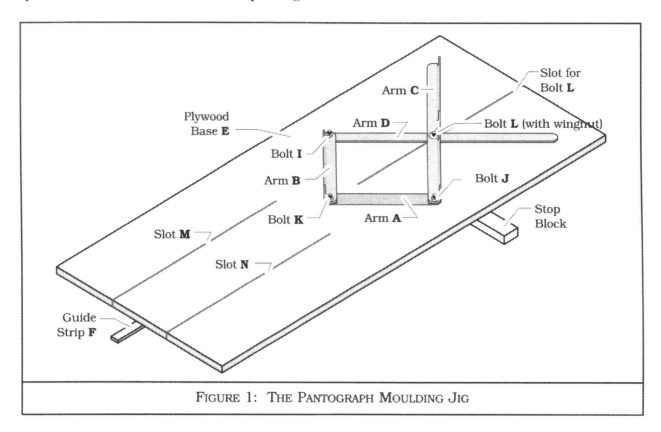

FIGURE 1: THE PANTOGRAPH MOULDING JIG

OPERATION

First, set a bevel gauge at the angle between the walls that are to receive moulding. Then, loosen the wingnut on the jig's bolt **L** and spread arms **C** and **D** wide enough that the gauge can lie between them. Finally, carefully slide bolt **L** forward, snugging arms **C** and **D** against the gauge, and tighten the wingnut. Completing a mitered corner requires two passes on the table saw. Set the jig's guide strip **F** in the saw's right-hand groove, and place a piece of moulding against arm **A**, so that it lies across slot **M**. Start the saw, and complete the cut by sliding the jig and moulding forward. To cut the companion piece, place the jig's guide strip in the saw's left-hand groove, place the piece of moulding against arm **B**, so that it lies across slot **N**, and repeat the sequence.

MATERIALS

Base **E** is a piece of $\frac{5}{8}$" plywood measuring about 1'-6" x 3'-6" — maneuverable and yet large enough to obtain a fairly wide range of angles. Guide strip **F** and the stop blocks can be made of pine or of hardwood. The jig also incorporates about 3' of angle iron and about 1'-6" of flat stock. Arms **A**, **B**, and **C** are made of $\frac{3}{4}$" or 1" angle iron. Arm **D**, which must be the same width as the angle iron arms, is made of flat stock to avoid interfering with the vertical wall of arm **C**.

CONSTRUCTION

1. Drill a hole for flat head bolt **K** in the center of the plywood base. The hole should fit the bolt snugly and be countersunk on the bottom of the base.

2. Cut the slot for bolt **L** in two router passes. First, using a straight bit slightly larger in diameter than bolt **L**, cut a slot through the base starting behind bolt **K** and extending rearward. Next, with a still larger bit, rout the bottom of the base to permit recessing the head of bolt **L**.

3. Cut guide strip **F**. The guide strip should be as long as the jig's base, narrow enough to slide easily in the groove of the table saw, and thin enough to allow the jig to lie flat.

4. Attach the guide strip to the bottom of the base parallel to the centerline of the plywood base and adjacent to the slot for the sliding bolt **L**, with countersunk wood screws.

5. Cut slots **M** and **N**. Place the guide strip in one groove of the table saw and cut a slot starting at the front of the base and extending halfway into the base. Repeat this process using the saw's other groove.

6. Cut the arms. Arms **A** and **B** should be about 9" long. Arms **C** and **D** should be left as long as practicable.

7. Drill a bolt hole near each end of arm **A** and near one end of arms **B**, **C**, and **D**. Using the distance between the pair of bolt holes in **A** as a reference, locate and drill a second bolt hole in each of the other arms. NOTE: This jig will not work if these four pairs of holes are not precisely the same distance apart.

8. Round off both ends of arms **A** and **B** and the bolt-hole end of arm **C**.

9. Provide wingnut clearance. Remove about 2" of the vertical wall near the bolt hole in the middle of arm **C**.

ASSEMBLY

1. Bolt **A** and **C** together with bolt **J**; and **B** and **D** together with bolt **I**. The arms should be snug but still able to move smoothly.

2. Insert bolt **K** through the hole in the base and attach arms **A** and **B** to bolt **K**.

3. Line up the remaining bolt holes in arms **C** and **D**, insert bolt **L** up through the slot in the base, and attach the arms to bolt **L** with a wingnut.

4. Attach the stop blocks to the bottom of the base. To determine their correct location, move the arms into their forwardmost position and tighten the wingnut. Place the jig on the saw, guided by either of the saw's grooves. Raise the saw blade to its maximum height, and without starting the saw, slide the jig forward until the blade nearly touches the jig's arm. With the jig in this location, hold the stop blocks against the rear edge of the saw's table and clamp them securely to the jig's base. Turn the jig over and screw the stop blocks to the base.

CONCLUSION

This jig is inexpensive and fairly easy to build. Moreover, the simplicity and accuracy of its operation make it quite useful, particularly on those shows that require miles and miles of moulding.

A System for Duplicating Tube Steel Frames

Geoffrey Webb

A problem often encountered while building scenery is the duplication of frames for identical units. Welding duplicate tube steel frames is doubly tricky because the distortion that occurs as the welds cool can easily pull a joint out of true. Single-purpose jigs like those typically nailed to a workbench do address the problem, of course; but they are not often intended to be reusable, and, once they're installed, space on the workbench is frequently unavailable for other projects. The use of duplicating clamps like the one illustrated in Figure 1 offers a better solution.

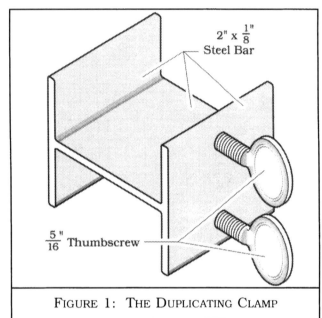

$2" \times \frac{1}{8}"$ Steel Bar

$\frac{5}{16}"$ Thumbscrew

FIGURE 1: THE DUPLICATING CLAMP

CONSTRUCTION AND USE

These clamps can be made quickly, easily, and economically from three pieces of 2" x 2" x $\frac{1}{8}"$ steel bar stock brazed together in an "H" section. As illustrated, two $\frac{5}{16}"$ thumbscrews fit into threaded holes on one side of each clamp.

After laying the frame to be copied on a bench or table, place clamps at various points along the members of this "master frame." Each member will require at least two clamps, and long members may require three or more. I have found it best to position the clamps 6" from the joints to be welded, to prevent their being covered with spatter and becoming hot enough to burn. In setting up, keep the thumbscrews conveniently out of your way on the inside of the master frame, and secure the clamps to the frame with finger pressure.

Now it is a simple process to lay precut lengths of tube steel in the clamps and tighten the remaining thumbscrews to hold all members securely in position during the welding and cooling of duplicate frames. After the accessible welds have cooled, the new frame can be unclamped, turned over, and re-secured in the clamps to permit the remaining welds to be made.

Such a clamp system provides a scenery maker with a versatile, portable, and economical tool. These simple clamps can be used in duplicating frames of virtually any shape or size and accommodate the most commonly used sizes of tube steel. Other sizes can be quickly manufactured as needed, of course, as can special-purpose clamps, such as those which might be of use in aligning chair legs to a chair base. The time spent constructing a series of clamps will be compensated for many times over in time saved setting up and adjusting.

❧❧❧

When a design called for ten swinging doors, the vises described below (common in the construction industry) enabled us to hold the doors on edge securely while quickly mortising them for hinges and strike plates. The vises' operation principle is simple: the weight of a door laid on edge on the ledger strip (**G** in the illustration below) activates a clamping action that holds the door firmly between movable jaw **F** and fixed jaw **C** as illustrated in Figure 1.

MATERIALS FOR A SINGLE VISE

> One 2'-0"-long 2x4; base **A**
> Two 1'-7$\frac{1}{2}$"-long; 1x4s nailing strip **B**; fixed jaw **C**
> Two 1'-7$\frac{1}{2}$"-long 2x4s; movable jaw **F** and movable jaw support **E**
> One 3$\frac{1}{2}$"-long 1x4; ledger strip **G**
> Four 3$\frac{1}{2}$"-long pieces of $\frac{1}{8}$" x $\frac{1}{2}$" steel strap; hinge **H** (two on each side)
> Two 3" x $\frac{1}{3}$" spiral springs **J** (one on each side)
> One 1'-4"-long piece of 1x ripped to 1$\frac{1}{2}$" wide; base stabilizer **K**
> Four $\frac{1}{4}$" plywood corner blocks with legs at 9$\frac{1}{2}$" and 1'-9" **D**

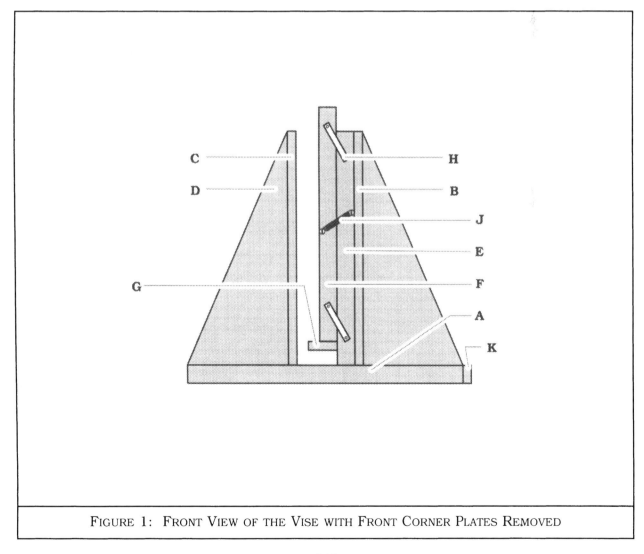

FIGURE 1: FRONT VIEW OF THE VISE WITH FRONT CORNER PLATES REMOVED

CONSTRUCTION NOTES (BUILD TWO)

It is useful to have a fairly heavy base for the vise — a 2x4 is appropriate. The 1x4 nailing strip and fixed jaw are held in position by the four $\frac{1}{4}$"-plywood corner blocks. The 2x4 movable jaw support reinforces the nailing strips, adds more mass to stabilize the unit, and provides a more convenient attachment point for one end of each of the $\frac{1}{8}$" strap steel hinges.

Each hinge measures $\frac{1}{2}$" x $3\frac{1}{2}$" and has a screw hole drilled at either end. One end of each hinge is screwed to the side of the movable jaw support; the other, to the movable jaw itself. The movable jaw's lower end is capped with a 1x4 ledger strip on which the door rests during use, causing the vise to close.

While many of the dimensions and particulars of construction of these vises can be modified to suit an individual shop's needs, it is important to insure that the hinges remain parallel to each other, with their pivot points aligned along the same vertical axes. Oversizing the holes drilled in the ends of the hinges will help insure proper alignment. During assembly, with the movable jaw held against its 2x4 support and 2" above the top surface of the base of the vise, attach the hinge plates at a 60° angle to the plane of the base. This, the jaw's open position, is maintained by two spiral springs, one mounted on each side of the movable jaw at 90° to the hinge plates. The springs reset the vise when the door is removed.

The final construction step is to add a $1\frac{1}{2}$"-wide piece of 1x cut to a length of 1'-4" to one end of each base, perpendicular to the axis of the base. This base stabilizer will prevent the vise from tipping while in use.

❧❧❧

Building scenery outdoors can be fun, but it has its problems. One of them is that it's usually not practical to drag the table saw or radial arm saw out of the shop, so one is stuck with using a portable circular saw to cut lumber to length. The cut-off jig shown in Figure 1 was used recently to help construct a setting that had to be built almost entirely on-site outdoors. It worked so well that it has been saved to be used on our main season indoors.

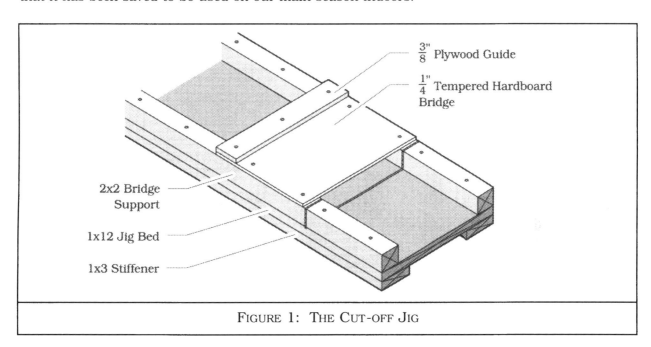

FIGURE 1: THE CUT-OFF JIG

The jig is constructed from a piece of 1x12 approximately 6' long. A 2x4 the same length is ripped in half to yield two 2x2s that are actually $1\frac{1}{2}$" x $1\frac{5}{8}$". These are attached to the top surface of the 1x12 parallel to each other, with their $1\frac{1}{2}$" faces on the 1x12. The bridge across the channel between the 2x2s is constructed from $\frac{1}{4}$" tempered hardboard. A strip of $\frac{5}{16}$" or $\frac{3}{8}$" plywood is attached to the top of the hardboard and acts as a guide for the edge of the sole plate of a circular saw. Care must be taken to ensure that the guide is mounted perpendicular to the 2x2s so that the jig will cut square. Two pieces of 1x3 are attached to the underside to stiffen the jig.

In use the jig is placed on two sawhorses or any other convenient support. The piece of lumber to be cut is first measured and marked with a small check, and then slid into position under the bridge of the jig. The circular saw is supported by the bridge and guided by the plywood strip as it cuts through the lumber.

The jig is almost as accurate as a radial arm saw for simple perpendicular cut-off work. The saw is easy to control so one hand can be used to hold the lumber in place. Unless the piece of lumber being cut is very long, the jig will adequately support the lumber so that no part breaks off or pops up as the cut is completed. The only problems with its use are that one must slide the lumber under the bridge from the end, and the space under the bridge is only $1\frac{5}{8}$": a warped 2x4 won't fit. The jig could be built with a larger space there, but a portable saw larger than 7" would have to be used. Even with these problems, it certainly beats the alternative of jostling sawhorses into position, kneeling on the board, and coaxing the saw to follow a pencil line.

An Inexpensive Pipe-Clamp Bench Vise

Patricia Bennett

While working on the setup of a new scene shop, our carpentry staff decided that a bench vise would be a useful addition to each carpenter's work station. The Technical Director and Assistant Technical Director priced various bench vises, all with 6" jaw widths, and discovered that they ranged from $75.00 to $150.00 each. An inexpensive and temporary alternative needed to be found until additional money became available for standard vises.

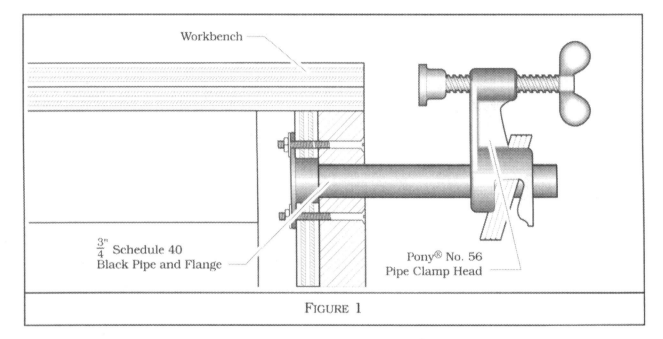

FIGURE 1

The problem was solved by the Shop Foreman, who showed us an inexpensive way to construct a bench vise of pipe clamps. Building the vise shown in Figure 1 requires only a workbench with a strong frame, a short piece of $\frac{3}{4}$" schedule 40 black pipe threaded at one end, a pipe flange, a plywood spacer ring, four bolts with washers, and a standard Pony® No. 56 pipe clamp head, the type of head that contains the extended multiple-disc clutch. The No. 56 head can be slid along the pipe, extending the length of the vise's throat to accommodate a range of widths. The total cost of the hardware is less than $20.00 per unit.

To construct the vise, drill a hole the same diameter as the pipe through the side frame of the workbench. Then thread the pipe into the flange and place the plywood spacer ring on top of the flange. Finally, slide this assembly through the hole from the inside of the table, and bolt it in place.

If long scenic units need to be clamped, such vises can be installed at both ends of the workbench. If heavy units are to be clamped, consider mounting the pipe flange on the inside of the opposite framing member and letting the pipe extend from the flange, across the width of the table, and through the hole in the vise-side framing member.

❧❧❧❧❧

One of the most versatile scenic, property, and costume construction products currently used in our industry is Ethafoam® rod. Uses for Ethafoam® rod are as numerous as the variety of designs presented onstage. Uses at Ohio State have included the crown piece for a dragon head, $\frac{1}{2}$" half-round moulding at the top of a Masonite® baseboard, $\frac{1}{4}$" leading for a prop stained-glass window, and simulated carvings on some chair backs. Available in diameter sizes from $\frac{1}{4}$" to 2" and beyond from both theatrical and commercial building suppliers, Ethafoam® has certainly proved its usefulness.

One problem often encountered when working with Ethafoam® is the need to split the round rod into half-rounds and quarter-rounds. When I first used Ethafoam® in properties construction at Iowa State, I used the rip fence of a standing bandsaw as a guide to split the rod. Control was difficult and safety was always a strong consideration. I had the good fortune of inheriting the Ethafoam® rod splitter described in this paper when I became the Technical Director at Ohio State. I wish I could take credit for its invention, but I cannot. I do wish to share the design of the marvelously simple device so that others do not have to reinvent the wheel.

The splitter is made from a 2" x 2" x 12" block of hardwood for the base, two $\frac{3}{4}$" x $\frac{1}{2}$" x $11\frac{1}{2}$" keeper blocks, 4 round head wood screws, and 5 standard utility knife blades. The version shown in Figure 1 has holes for $\frac{1}{2}$", $\frac{5}{8}$", $\frac{3}{4}$", 1", and $1\frac{1}{4}$" rod.

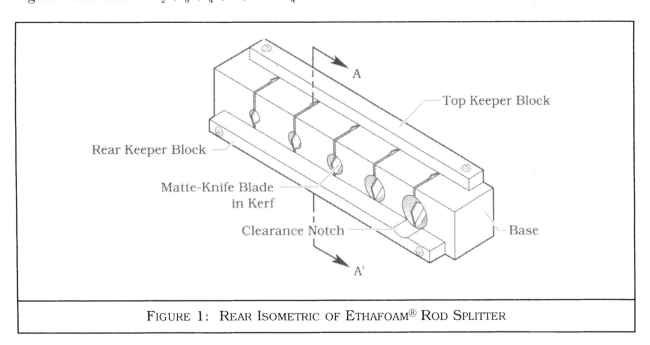

FIGURE 1: REAR ISOMETRIC OF ETHAFOAM® ROD SPLITTER

Begin construction of the base by drawing a centerline down one face of the hardwood block. Then, using spade bits, drill appropriately sized holes on 2" centers along that center line. Next use a coping or band saw to cut a saw kerf along the center of each of the drilled holes. The kerfs, which will serve as slots for holding matte knife blades in place in the holes, should start at the rear of the base and extend forward at about 45° for good results. See Figure 2.

Once all the kerfs have been cut, the keeper blocks should be placed as indicated in the drawing. Pre-drill the holes for the hold down screws and carve away the small portions of the rear keeper block that interfere with any of the holes.

The last step in making the splitter is to install the utility knife blades in the saw kerfs and screw the keeper blocks in place.

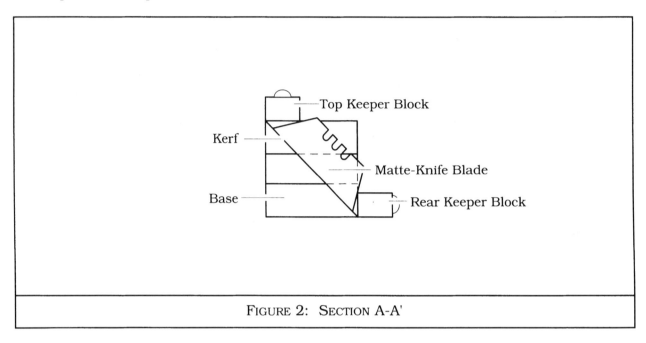

FIGURE 2: SECTION A-A'

Pulling the Ethafoam® rod through the appropriate hole and across the utility knife blade produces half-round stock. For quarter-rounds, pull the rod through the splitter a second time. For some of the smaller diameters, start the split using a blade in one of the larger holes, then feed the stock through the appropriate hole. Dry lubricant can aid in the smooth operation of the Ethafoam® splitter when making large quantities of stock.

TECHNICAL BRIEF

Sound

Using a Piano to Create a Reverberation Effect

Serge Ossorguine

The technique described here provides the Sound Designer with an alternative means of creating a reverberation effect. This technique utilizes the phenomenon of sympathetic vibrations induced by sound waves traveling across the strings of a piano. This phenomenon can be demonstrated by depressing the sustain pedal of a piano and speaking loudly (or making any loud noise) into the strings of the piano. The strings will respond by vibrating at the frequencies of the source sound. The effect is an apparent reverberation of the source sound in a large concert hall. The reverberation time, depending on the amplitude of the source sound, can be up to four seconds.

Provided there is a piano available in an acoustically isolated room, the cost of assembling this device is considerably less than any commercially available reverberation device of reasonable quality. Besides economy, the technique described here offers a wide range of both tonal and reverberation time control.

SETUP

In order to set up a system where you can feed a variety of source signals into the strings and monitor the reverberant signal, you will need the following items:

> Piano (any size)
> Full-range loudspeaker with amplifier
> Contact microphone (one or more)
> Multichannel audio mixer (minimum 2 channel)
> Block of wood

1. Block the sustain pedal (the one on the right) in the open position by depressing the pedal and placing the block of wood underneath the push-rod.

2. Place the speaker facing the strings of the piano. As a general rule, the speaker should be placed so as to optimize the strings, sympathetic vibration. On a grand piano, open the lid and point the speaker perpendicular to the length of the strings. On an upright, open the lid underneath the keyboard and point the speaker directly at the strings.

3. Place the contact microphone(s) on the soundboard (the wooden panel behind the strings) near the higher registers. When using more than one mic, space out the placement on the soundboard. Keep in mind that once into the tone selection procedures (see below), the mic placement may need to change.

4. Make connections as shown in Figure 1.

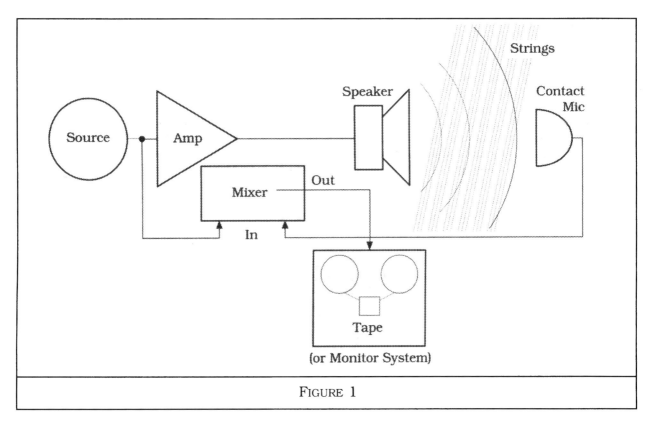

FIGURE 1

ADJUSTMENT

Once all the connections are made, you may want to adjust the system. As in any reverberation device, the ability to control the amount of reverberation time and a means of selecting a tonal quality are essential in rendering this system useful to the Sound Designer. There are basically two ways one can adjust the reverberation time of this device. The simpler way is to mix the source signal higher than the reverberant signal. The subtler way is to place pieces of cloth on the strings. As for tonal quality, one can explore the variety of harmonic qualities of the source sound as it travels across the body of the piano by moving the mic(s) to different areas of the piano. For example, by placing the mic on the steel frame rather than the soundboard, the resulting reverberant signal will be completely different.

CONCLUSION

The piano is a musical instrument. Using a piano to create a reverberation effect is simply another way of playing it. Just as there are many interpretations of a Beethoven sonata, there are many ways of arranging this system to create reverb. One needn't wear a tuxedo, but donning the patience and ear of a musician is helpful in working with this system.

෴෴෴

A Versatile Audio Power Meter

Steven E. Monsey

Many plays involve the use of practicals like radios and TVs whose speakers must be driven by one of the theatre's system amplifiers. This condition poses a problem, for system amplifiers can easily overdrive and damage speakers that are small enough to be mounted in practicals. And, unfortunately, most theatrical sound systems do not contain a meter that permits a sound board operator to monitor the health of a given amplifier/speaker combination.

The audio power meter described in this article is designed to fill that gap. Because this unit monitors the interaction between so many amplifier output levels (0.1W to 400W) and speaker sizes (4Ω, 8Ω, and 16Ω), it can be used with a large number of combinations.

DESCRIPTION AND CONSTRUCTION

Any metal or plastic box works nicely as a housing for the unit. Terminals for speaker connections should be mounted on the outside of the housing to simplify hookup.

As designed, the power meter includes two rotary switches: one to be set at the amplifier's output wattage; the other, at the speaker's load impedance. The unit's heart is an integrated-circuit chip that senses changes in voltage levels and drives a set of ten LEDs, providing a visual display of system performance. To permit easy reading of the display, the unit also contains a transistor circuit that slows down the occurrence of the peak voltages that turn the LEDs on. The meter, powered by batteries or by a 12V power supply, draws about 2.5W when all the LEDs are lit.

COMPONENTS

Resistors (All are $\frac{1}{4}$W.)

R1:	10KΩ
R2:	1KΩ
R3:	180KΩ
R4:	1MΩ
R5:	2.7KΩ
R6:	390Ω
R7, R9:	3.9KΩ
R8, R10, R12:	10KΩ
R11, R13, R15:	18KΩ
R14, R16, R18:	30KΩ
R17, R19, R21:	47KΩ
R20, R22:	68KΩ
R23:	100KΩ

Other Components

- Q1: 2N3906 PNP Transistor
- S1: 3-position Rotary Switch
- D1: 1N194 Diode
- S2: 6-position Rotary Switch
- C1, C2: 2.2μF, 16V Tantalum
- LEDs: Any commonly available LEDs will work.
- IC: LM3915 LED Dot/Bar Display Driver (Additional information on the LM3915 can be found in the *National Semiconductor Linear Handbook*, 1982 ed.)

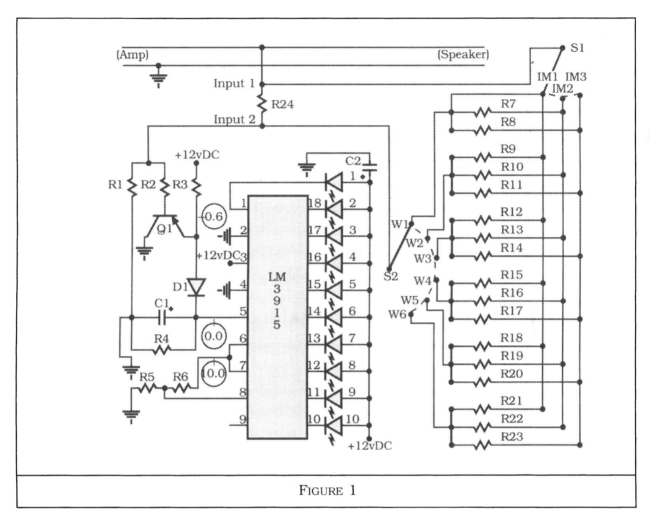

FIGURE 1

ASSEMBLY NOTES

1. Use alligator clips as heat sinks during soldering.

2. To dampen objectionable noise from the LM3915, the wires leading to the LEDs should be no longer than 6", and capacitor C2 should be wired as close to the LEDs as possible.

3. Install the components in the following order: resistors, capacitors, diodes, transistors, and finally, the integrated chip.

4. Observe polarities in installing all capacitors, diodes, transistors and the chip. Use an IC socket with the chip.

5. If you are using rotary switches, connect the S1 common at Input 1 and the S2 common at Input 2.

6. Connect the speaker's ground and positive leads to the speaker terminals.

7. Label the LEDs with the appropriate wattage rating.

OPERATION

Connect the unit parallel with the amp. Set switch S1 at the impedance of the speaker being driven (4Ω, 8Ω, or 16Ω) and switch S2 at the output wattage of the amp (0.1 to 400W), and turn the unit on. Some of the LEDs should light up whenever the amp is providing a signal. If they do not, turn the amplifier off, check fuses, and check for correct voltages at the test points shown in Figure 1. Once the system is working properly, the Sound Board Operator protects the speaker by turning the amplifier output down whenever all the LEDs are lit.

₰₰₰₰

Occasionally, a production requires the sound of explosions on or near the stage. Recorded explosions are fairly inexpensive and undoubtedly safe, but a genuine explosion has a greater effect on an audience because it is experienced firsthand. Using a concussion mortar is one way to produce a live explosion effect.

A concussion mortar can be purchased from a manufacturer for $60 to $70. For about $15, however, any technician familiar with a drill press and metal lathe can make the mortar shown in Figure 1 in about three hours. Further, during manufacture the technician can tailor the mortar to produce explosions that will meet fairly specific sound design criteria.

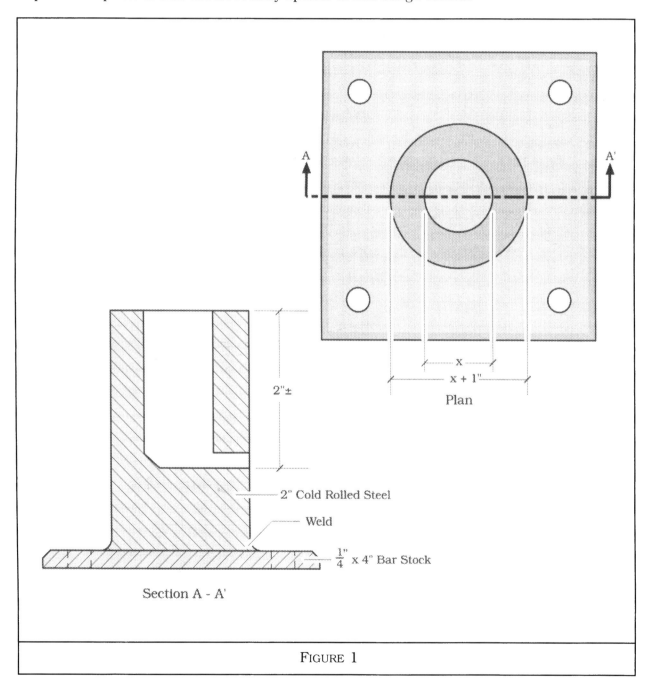

FIGURE 1

Notes on Designing and Building a Concussion Mortar   *David C. Bell*

TAILORING TONAL QUALITY AND VOLUME

A mortar's tonal quality is directly related to the diameter and the depth of its bore. Shallow, narrow bores should produce sharp, report-like explosions. The larger the diameter of the bore, the deeper the pitch; and the deeper the bore, the "hollower" it will sound. The intensity or volume of the explosion a mortar produces is determined by the size of the charge.

The accompanying drawings illustrate a typical concussion mortar. The model illustrated here was made of a 6"-long piece of 2" cold rolled bar stock welded to a base plate of $\frac{1}{4}$" steel plate. Charged with one-third ounce of two-component concussive powder, it produced a relatively deep bang. Quarter-ounce and half-ounce charges resulted in explosions of different volumes but similar tonal qualities.

CONSTRUCTION

A few cautions and suggestions apply to all mortars:

1. Use cold rolled round steel bar in preference to a piece of schedule 40 or schedule 80 black pipe. The force of the explosion could rupture a pipe seam, creating a hazardous situation.

2. Make the bore at least 2" deep with a minimum wall thickness of $\frac{1}{2}$".

3. The size of electric matches varies from manufacturer to manufacturer. Determine the diameter of the match hole by measuring the match to be used.

4. Design the steel base plate so that it provides stability and permits secure attachment to a floor.

SAFETY NOTE

State and/or local regulations in some areas may prevent anyone other than a licensed pyrotechnician from setting off concussion mortars. In any event such devices should be handled carefully, following the most stringent safety precautions.

੨੩੨੩੨੩

An inexpensive electret microphone element can make a good substitute for the higher-quality dynamic microphone usually used as the pickup mic in a stage monitor system. The electret's sensitivity is excellent, and its noise level low enough that the expensive dynamic mic can be freed for use elsewhere. Furthermore, if you don't have your monitor mic caged against theft, use of an electret like the one described below reduces the probability that a thief will decide to take it.

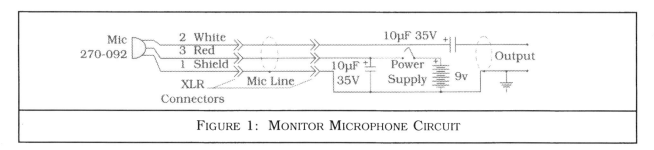

FIGURE 1: MONITOR MICROPHONE CIRCUIT

Figure 1 shows the microphone circuit we used at Michigan Tech. Bill Isaacson put it together out of parts on hand, so the cost was very low. The only thing we had to purchase was the Radio Shack mic element, which cost $2.69. Bill mounted the mic element directly to the cord end of an A3M connector with a few turns of electrical tape. The mic dangles just below the catwalk of our black-box theatre at the end of a short piece of microphone cable. One has to look at it twice to realize that there is anything hanging there other than a mic cable.

The mic feeds the microphone input of a McMartin 108C amplifier. The amp's input is a screw terminal strip designed for a balanced input. We hooked up the unbalanced output from this microphone by tying one of the balanced input screws on the amp to ground, and we've had no problems from excessive noise pickup with the unbalanced microphone line.

The mic element contains a small pre-amp/impedance converter that needs from 2 to 10 volts to operate. We used a 9V battery for power because we had them on hand. Our wireless headset system eats 9V batteries, but a battery that will no longer supply the current necessary for a headset to transmit will still provide the 1mA needed to run the mic element. We always have a good supply of used batteries that will run our monitor mic. A few penlight batteries or a small AC adapter for a cassette recorder could be used instead.

The omnidirectional pickup pattern of the mic element works well in our black box. A proscenium theatre will frequently have a cardioid mic mounted in the first beam position and pointed at the stage for a monitor mic. If you want to replace one of these with an omnidirectional electret element, you may want to move the mic to someplace above the stage to avoid picking up too much audience noise. The mic is not likely to be damaged because it is so small, light, and rugged, but if somehow you manage to completely smash it with flying scenery you won't have lost much: the microphone and connector cost less than $6.00.

PARTS LIST

1	9V Battery Connector	1	Electret Mic Element, Radio Shack 270-092
2	10μF 35V Capacitors	1	A3M Connector
1	SPST Switch	1	A3F Connector
	Shielded Wire	1	9V Battery

Frequently, when microphones are used to reinforce stage performances, sudden rises in volume overdrive audio equipment causing both signal distortion and audience discomfort. In an instance such as this, a compressor/limiter may be used to control the output signal level, thereby protecting amplifiers, speakers, and ears. This article details a simple circuit that provides quality audio compression/limiting at low cost.

COMPRESSION AND LIMITING

An audio compressor is a device that attenuates a signal once it reaches a set threshold voltage (Vth), in effect turning the gain down automatically when the input signal gets too large. How much a compressor attenuates a signal is defined by the limiting ratio, *i.e.*, the ratio of the rise in input level to output level. The higher the limiting ratio, the more the audio signal is compressed. In the extreme case where the limiting ratio is infinite, the circuit is said to be "hard-limiting" and there is no rise in output level past the threshold voltage.

The circuit, shown in Figure 1, can be constructed from parts easily obtainable from any electronics supply store: five resistors (100KΩ, 100KΩ, 10KΩ, 10KΩ, and 556Ω); two 10KΩ variable resistors, two standard op-amps, and three diodes. In addition to these elements, a DC voltage supply of 15V, or the equivalent, is required for power.

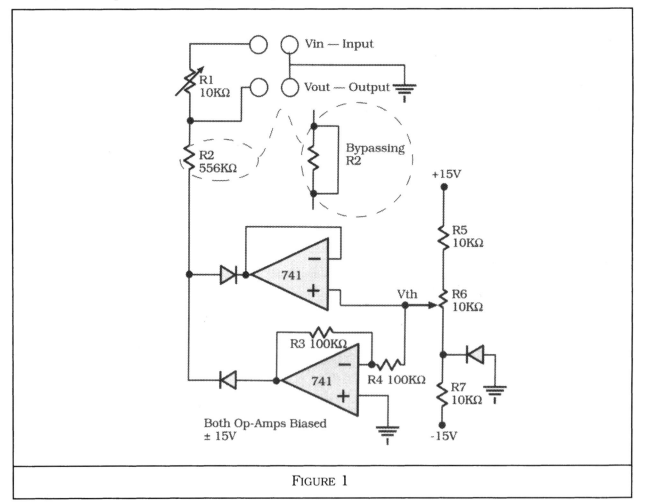

FIGURE 1

The variable resistors R1 and R6 are used to set the limiting ratio and the threshold voltage, respectively. The limiting ratio is defined as (R1 + R2) 4 R2 : 1. Therefore, if R2 is bypassed, the limiting ratio becomes infinite and this circuit hard-limits at Vth, the threshold voltage. The threshold voltage can be set by measuring Vth with a volt-meter while varying R6. As designed, this circuit's limiting ratio can be varied from 1:1 to 20:1 (or simply be set to hard-limit) with a Vth that is variable from 0V to 8V.

LEDs may be used for the diodes as long as all three diodes are of the same type (or at least have the same voltage drop across them.) Also, setting R1 = 0 ties the input directly to the output and effectively bypasses the circuit.

This compressor/limiter circuit was originally designed and tested by Joseph Rimstidt.

¿▲¿▲¿▲

Horn-Hat Mics for Sound Reinforcement

Tien-Tsung Ma

Sound reinforcement often requires expensive microphones such as PZMs, shotguns, or wireless-es. Whenever a production's sound budget does not allow for the purchase or rental of such equipment, the use of a shop-built "horn-hat" microphone can solve the financial problem and will provide satisfactory results.

Horn-hat mics like that shown in Figure 1 are made by bolting a speaker horn and a microphone together. Speaker horns, which are designed to diffuse sound waves evenly, work just as effectively as sound-wave collectors. Thus, the addition of a speaker horn "hat" to a microphone allows the mic to pick up sounds that originate within a broad area, the size of which is dependent on the sound-diffusing angle of the horn. Though not commercially available, horn-hat mics have been used for years to pick up vocals onstage. Their use as sound reinforcement tools, however, has only recently been explored.

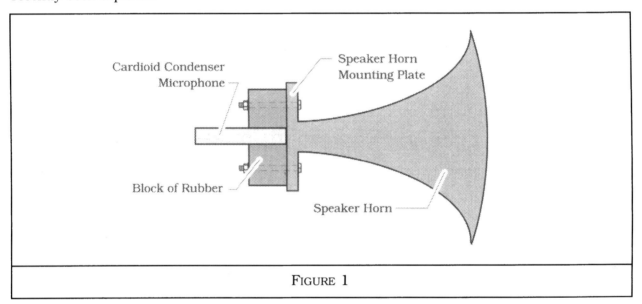

FIGURE 1

CONSTRUCTION

Materials

>   Speaker Horn
>   Cardioid Condenser Microphone
>   Block of Rubber (available from McMaster-Carr and others)
>   Nuts, Washers, and Bolts

Steps

1. Cut the rubber to $1\frac{1}{2}$" thick.

2. Drill a snug-fitting hole through the rubber for the mic.

3. Drill bolt holes through the rubber.

4. Bolt the rubber and the speaker horn together.

5. Insert the cardioid condenser mic into the rubber so that the head of the mic and the bottom edge of the speaker horn are flush.

COMMENTS

Horn-hat mics are larger and perhaps more awkward to hang than most microphones, but their sound-capturing qualities are comparable to those that are commercially available and much more expensive. Both speaker horns and cardioid condenser mics are common equipment in most theatres, and assembly costs are minimal. The horn-hat mic is more efficient than most shotgun mics and regular cardioid mics because of its wider sound-collecting angle; and carefully designed horn-hat mics can restrict unwanted sound better than PZMs and regular cardioid mics.

NOTES

1. A horn with a wider sound-diffusing angle can cover a larger area than one with a narrow sound-diffusing angle, but as a guide, the combination of a Beyer M201 mic and an EV HP640 horn has been used to cover a 60' x 40' area onstage with sound loss limited to less than 6dB.

2. A cardioid dynamic transducer can be used in place of a cardioid condenser mic if such a mic is not available, but such a substitution results in a less efficient device.

3. Like other mics, horn-hats require the use of an equalizer to modulate the frequency response.

4. Different mic/horn combinations will result in different sound characteristics. The frequency response of the mic and the sound-diffusing angle of the horn are the two factors that determine the characteristics of any horn-hat mic.

5. One horn-hat mic can cover the same amount of area as several shotgun mics without producing phase problems.

᪥᪥᪥᪥

An Audio Test-Tone Generator

Jim van Bergen

Before each performance in a lengthy run or repertory situation, the signals of any microphones to be used must be set at predetermined levels. Commonly, during this part of the sound check, the sound operator makes the necessary adjustments at the mixing console while an assistant approaches each mic in turn and repeats the familiar "Testing: one, two, three." The results of such mic tests are often unsatisfactory because of inconsistencies in signal gain, distance from the mic, and the uniqueness of each assistant's voice. Worse yet, damage can occur when a novice assistant resorts to blowing or tapping or scratching on the mics. The test-tone generator illustrated in Figure 1 eliminates many of the variables in sound checks: it provides common frequencies and a standard gain every time it is used. The circuit on which it is based appears in *Getting Started in Electronics* by Forrest M. Mims, III, which is available at Radio Shack stores.

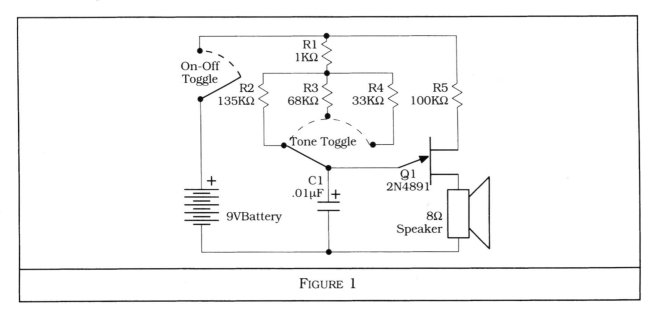

FIGURE 1

I have adapted his design by replacing a variable resistor with a three-position tone toggle, and by adding a DC power supply and a switch to supply discrete and constant tone signals. This efficient and inexpensive circuit can be built quickly from readily obtained parts. It is compact and will fit nicely into a 4" x 3" x 1" or smaller project box, depending on the size of the full-range speaker used. Adding a belt clip or neck cord allows for no-hands operation and lets the assistant watch the sound operator rather than the mic.

The tone frequency is determined by how fast the capacitor charges through the resistors, and more experienced circuit builders may want to substitute other resistors for those indicated here. The recommended 135KΩ, 68KΩ, and 33KΩ resistors, however, adequately simulate the dominant human speech frequencies for both sexes by producing tones around 600Hz, 1200Hz, and 2400Hz, respectively.

Using the generator is an easy process. Set the toggle to the correct frequency, hold the generator about 6" from the microphone to be tested with the generator's speaker aimed at the mic, and switch the unit on. A novice can easily operate this important instrument, allowing technicians the luxury of using a different assistant on each check if necessary. Since this device provides identical sound pressure levels at each of two frequencies each time it is used, its use promises superior results.

Combined Topical Index: Volumes I and II

Rigging Techniques

Safety

Scenery

Scenery Electronics

SCENERY MECHANICS

SCENERY TOOLS

SOUND

Printed in the United States
by Baker & Taylor Publisher Services